ZOOM

ALSO BY BOB BERMAN

The Sun's Heartbeat

Biocentrism (with Robert Lanza, MD)

Shooting for the Moon

Strange Universe

Cosmic Adventure

Secrets of the Night Sky

ZOOM

HOW EVERYTHING MOVES

FROM ATOMS AND GALAXIES
TO BLIZZARDS AND BEES

BOB BERMAN

LITTLE, BROWN AND COMPANY

New York   Boston   London

Little, Brown and Company
Hachette Book Group
237 Park Avenue, New York, NY 10017
littlebrown.com

First Edition: June 2014

Little, Brown and Company is a division of Hachette Book Group, Inc.
The Little, Brown name and logo are trademarks of Hachette Book Group, Inc.

The publisher is not responsible for websites (or their content) that are not owned by the publisher.

Unless otherwise noted, all photographs are by the author.

The Hachette Speakers Bureau provides a wide range of authors for speaking events. To find out more, go to hachettespeakersbureau.com or call (866) 376-6591.

ISBN 978-0-316-21740-8

LCCN 2014937386

10 9 8 7 6 5 4 3 2 1

RRD-C

Printed in the United States of America

To the memory of my mom, Paula Dunn

Contents

CONTENTS

ZOOM

"Forward"

The heavens rejoice in motion...
—JOHN DONNE, *ELEGIES* (CA. 1590)

We are embedded in a magical matrix of continuous motion. Clouds change shape, tsunamis destroy cities. Nature's animation happens eternally. Its energy springs from no apparent source. Nor, we learn, does it ever diminish. It's tireless.

As we have with all magic, we've grown accustomed to nature's endless guises. Too accustomed; we scarcely give it a second thought. Yet it is intimately close. Even the workings of our eyes and brains, reading these words, are examples of natural motion. In our minds' case it's the action of electrons and neurons as one hundred millivolts of electricity make various connections in the brain's one hundred trillion synapses. The result: our perceptions.

This book, then, is about natural activity in all its forms. It is essentially a book of miracles. To paint this dynamism in the vivid colors it deserves, I will offer close-up peeks at the most fantastic, epic, intriguing, but also little-known ways in which spontaneous action operates, using the discoveries of scientists from ancient times to the twenty-first century.

Because motion is everywhere and takes all forms, this cannot possibly be an exhaustive survey, though I have endeavored to include all nature's major theaters, such as wind, digestion, and shifting poles.

A mere dry recital of facts and data wouldn't be much fun. So let's marvel—not at man-made motion, even if our rockets and bullet trains are indeed wonderful, but at the kind that unfolds on its own. This book itself moves, too, after the opening high-speed salvo, from the slowest entities to the fastest ones. Along the way I've paused to recount the stories of some of the fascinating people who brought us discoveries in various venues. Some were geniuses. Others were lucky. Many were so far ahead of their time they were ridiculed.

This, then, is our story—of the endless movements that forever surround us and the brilliant people who uncovered these revelations through the centuries. And how destiny's own quirky momentum carried them through their lives.

Bob Berman
Willow, New York

Damage and Escape

It's a warm wind, the west wind,
full of birds' cries...
— JOHN MASEFIELD, "THE WEST WIND" (1902)

The storm was scary-wild.

Although it had lost hurricane strength before slamming into upstate New York, the wind still howled at fifty-five miles per hour, and the dog hid under the bed. But it was the rain, the relentless rain, that had us all worried. By the second day, more than eight inches had fallen. In our mountainous area, streams overflowed before the first sunrise. Many wooden spans, along with two steel-and-concrete bridges, did not survive the night. They were simply gone, vanished without a trace. Authorities later assumed they must be lying at the bottom of the enormous reservoir twenty miles downstream.

Entire communities were isolated from the world. At noon that day, some homes, those still inhabiting their footprints, had water up to their windowsills. Meanwhile the ground had become so soft and wet that the gales had no trouble knocking down swaths of trees, root balls and all.

The power went out the first night. In our rural region, which doesn't offer mail delivery or cell-phone service even on the sunniest days, we were utterly alone. No one had water, plumbing, or telephones. It might as well have been the year 1500.

Morning dawned to find trees across my roof. Shattered glass littered the stone entranceway. But this wind-borne destruction paled next to the devastation wrought by the waters moving through those valleys. My niece lost her entire house. It had been standing placidly for forty years, and then it was gone. The flood-waters had been five feet deep and had crept along at less than four miles per hour. Yet this sluggish brown water had created far more devastation than my own backyard's fifty-five-mile-per-hour gales.

It was ironic, in a way. For decades I had made my living narrating nature's activity as though I were a sportscaster. As the astronomy editor of the *Old Farmer's Almanac* and a columnist for *Discover* and then *Astronomy* magazines, I would routinely calculate how the moon and planets moved and describe their exuberant conjunctions. Nature's motions reliably put bread on my table. Now they had turned on me. Like everyone else, I wondered how many out-of-pocket repair dollars I would have to spend.

If nature's activity had long paid my mortgage and yet was now chasing me out of the house, I smelled a story whose dramas, tragi-comedies, and linked biographies might equal any novel's. I already knew that water moving at just four miles per hour is as destructive as a medium-strength tornado, which is why floods kill more people than windstorms. Water is eight hundred times denser than air and pushes things far more easily. But who were the first scientists to discover this? Were they impelled by personal events, as I was? Did their own lives and struggles include dramatic episodes?

Nature's whimsy obsessed me even before the winds quieted. I realized that my ideas about physical and biological animation were themselves forms of motion on the neural-electrical level. So it's *all* motion, everything of interest, always.

After the storm, the power remained out and would not be restored for more than a week. I scribbled notes by hand, by candle-light, the way Thomas Jefferson had done, for he was obsessed with natural science, too. (John F. Kennedy, presiding over a 1962 gathering of forty-nine Nobel laureates, commented that never before had such talent been assembled at the White House, "with the possible exception of when Thomas Jefferson dined alone.")

My lifelong preoccupation with celestial motion then expanded to the movements of desert sand, disease, and maple sap. As the heavy rain continued, I recalled that it falls at twenty-two miles per hour. In nature, this same number repeats over and over, like the restatement of a musical motif. Were there other recurring patterns? What were the boundaries—the fastest and slowest in the universe and in everyday life?

I knew I would have time on my hands while workmen repaired the house. I made my decision. I would raid my savings and travel the world. I'd use my press credentials to find the experts and researchers who probe the most amazing motions of nature. I would explore and expose anything that stirs, budges, or animates itself, from the strangest and slowest entities to the very fastest. I would also research the ways in which earlier cultures uncovered these secrets.

One adventure had barely ended. A much bigger one was beginning.

And I knew exactly where I would go first.

PART I

FIGURING OUT THE
MOTION PICTURE

CHAPTER 1: *The Growth of Nothingness*

Journeys in an Exploding Universe

And what does this awesome motion mean?
—NIKOLAI GOGOL, *DEAD SOULS* (1842)

From an observatory atop a mountain in the Andes one moonless midnight, untwinkling stars peppered the heavens. Not even the smallest patch of celestial real estate was starless. The Milky Way split the universe in two with such ferocious brilliance that the observatory's giant domes cast blurry, surreal shadows on the ground.

The clank of footsteps on the building's metal catwalk broke the silence. It was the observatory director, Miguel Roth. He stopped and casually surveyed the scene as if he hadn't spent the last twenty years in this place. Handsome as a movie star, Roth is the undisputed godfather of the researchers who live in this thin air, breathing in the cosmos from the most perfect astronomical site on earth. That night Dr. Roth was generously accompanying a lucky visiting American journalist interested in how the world moves.

That was me.

Looking for motion, I had gone for broke.

Starting at my local public library, where things move with a wonderful slowness, I had hunted for the earliest recorded thoughts about nature's animation. I toyed with the logical approach of beginning with viruses, fingernails, or tectonic plates—things that move and grow so slowly, so grudgingly, that they are imperceptible. Start slow and build from there.

But action films never begin with lethargy. I liked the idea of moving from slow to fast, but why not shoot an opening scene in which *everything* hurtles at screaming velocities? Hit the ground running with frantic exploits only the lunatic mind of nature could choreograph? After all, the biggest of all known motions encompass everything we could possibly consider.

But that hysterical realm is not of this earth. The mother of all motion is the entire universe, which is blowing itself apart. As it does so it creates separate animated venues, like swirling eddies in rapids.

Venture off the planet, and wildness rules. The smallest solar telescope shows that even on our nearby sun, that beloved giver of life, it's always the apocalypse. By the time we gaze toward the farthest galaxies, we see spinning, colliding things that are tumbling faster than the speed of light.

But how can mortals even perceive a universe that is exploding around them? I needed to visit cutting-edge astrophysicists working with the world's best equipment, people who are used to thinking outside the box. And, moreover, that "box" is not some well-behaved container but rather the spherical earth, crazily hurtling like jettisoned cargo toward no particular destination. No one's built a major observatory east of the Mississippi for more than a century, so this meant a distant odyssey. The problem is not just clouds. Astronomers require steady "seeing" (nonblurry images), and this happens when overhead air is spared the turbulence of multiple temperature layers. Mountaintops are good, but the most ideal locations do not exist in the mainland United States or Europe—not even in Asia's Himalayas. They are in South America. The reason for that continent's top celestial status has odd roots, entangled with a long-dead Scotsman.

The Scotsman was industrialist Andrew Carnegie, and he was easy to hate. While his workers lived from hand to mouth, unsuccessfully fighting their boss's miserly pay cuts, he became the world's richest man. By the late 1800s, Carnegie Steel Company—later called United States Steel Corporation—propelled its tiny owner, exactly five feet tall, into a royal life in a Scottish castle.

But everyone loves a converted tyrant, a sinner turned saint. Carnegie did a 180-degree turnaround just as the century flipped from the nineteenth to the twentieth. In a series of newspaper articles, the robber baron, outperforming even the metamorphosis of Dickens's Ebenezer Scrooge, started advocating the abolition of war and free nonsectarian education. Donating fantastic sums that ultimately totaled his entire fortune of $380 million—several billion in today's dollars—he established more than three thousand free libraries, funded African-American education, built concert venues (Carnegie Hall comes to mind), and (in case you thought this was never heading toward science) established a series of cutting-edge foundations. Thus came the birth of the Carnegie Observatories, a unique institution that still is working full-time on the greatest mysteries of the universe, which, fortunately for us, revolve around motion on the most epic scale.

Carnegie hired the best possible person to be the first director of his fledgling institute—George Ellery Hale, who in turn rounded up the sharpest minds of his time. First he employed the renowned Harlow Shapley, the person who found that Earth does not sit inertly at the center of our galaxy like Jabba the Hutt. This was the biggest motion-related headline of the newly born twentieth century. The sun and Earth, he found, lie closer to the edge of the galaxy than the middle, and thus they whirl around as it spins.

Next Carnegie hired Edwin Hubble, freshly returned from studying in Oxford, where he acquired a British accent that, annoyingly, he never shed. It drove his colleagues bonkers.

Hale and Carnegie believed that great discoveries demanded

the world's biggest telescope, and they set out to build it. Site tests for the area around Mount Wilson, then a dark and sleepy region outside Los Angeles, began in 1903. The men soon completed a behemoth that had a mirror sixty inches wide, the largest in the world. Then in 1917, also at Mount Wilson, they outdid themselves with the completion of the hundred-inch Hooker colossus and its nine-thousand-pound optical surface made of melted wine-bottle glass, which explains why that telescope has a green mirror, a fact capable of stumping any *Jeopardy!* champion. In that era before rural electrification, each telescope precisely tracked stars with a mechanical drive mechanism propelled by two-ton falling weights.

It was there that Hubble, haughty and unpleasant but one of the century's best observational astronomers, photographed a special type of variable star and concluded that a famous bright elliptical blob in the constellation Andromeda is not merely a nearby nebula but a separate "island universe"—a remote empire of billions of suns. All the other spiral nebulae, he reasoned, must likewise be independent starry kingdoms stretching off into the distance. Instantly the cosmos became a million times larger.[1]

When I phoned Carnegie Observatories' current director, Wendy Freedman, she told me that the discovery "ranked with the Copernican revolution."

"Yes, Edwin Hubble may have been arrogant," she conceded. "But there are only a few times in history when you change the very nature of the universe. You can't take that away from him."

Nor could you take credit away from the young Carnegie Institution for Science, which conjured milestone after milestone as though they were cards from a magician's sleeve. Director Hale personally created the National Academy of Sciences. His astronomers announced that elliptical galaxies have only old stars, whereas spirals are still making new ones. These were bombshells. But the biggest of all was the discovery that is relevant to our pursuit: the 1929 announcement that the universe is expanding.

This had never been suspected. No holy book of any religion, no Renaissance scientist, no philosopher had ever written that the entire cosmos is growing larger. Indeed the early Greeks, superb logicians that they were, would have no doubt dismissed that notion as meaningless. If everything expands at the same time, how could anyone possibly know it was happening?[2]

The first hint that the cosmos is squirmy had arrived in Einstein's brain in 1915, when he crafted his theory of general relativity because his math just would not work in a static universe. Yet the universe was assumed to be stationary—it was a "given," a truism, and Einstein had no reason to doubt it, so he famously added a fudge-factor number that he called the *cosmological constant*. Thereafter, his equations worked just fine. But when Hubble found that virtually every galaxy displays a redshift, indicating its rapid recession from us, the conclusion was inescapable: the universe is blowing up. Galaxy clusters are separating from their neighbors. Einstein had predicted it within his mind without peering through a single telescope, and he would have announced it if only he had more confidence in himself. "The worst blunder of my life," he famously muttered to anyone who would listen.

Here was motion on a scale of which no one had dreamed. Even nearby galaxies, those that live within the closest one thousandth of one percent of the cosmic inventory, which are the slowest moving of them all, rush away from us at a speed of 1,400 miles per second. Those dwelling at a "mere" one billion light-years from Earth zoom away at 14,000 miles per second. That's 28,000 times faster than a high-speed bullet.

The visible stars that fill the night sky *can't* move faster than 600 miles per second or they would escape the gravitational clutches of the Milky Way, never to return. Speeds like 1,400 miles per second—meaning you could go from Coney Island to Hollywood in the time it takes to say, "Got your seat belt on, honey?"—were bewildering in 1929. Yet such motion was couch-potato leisurely compared to what the next-gen postwar telescopes would reveal soon enough.

The newly observed speeds were breathtaking and also dispiriting. It became obvious—and remains so today—that no matter what system of propulsion is invented in the future, we will never visit the vast majority of galaxies: they are fleeing faster than we can ever hope to approach them.

The Sombrero galaxy, containing two hundred billion stars, rushes away from us at 562 miles a second. *(Matt Francis)*

Hale, still not finished despite suffering years of serious health problems (to which he finally succumbed in 1938), raised money for the next colossus, the two-hundred-inch telescope on Palomar Mountain in Southern California. It opened in 1949 and had a light-gathering mirror as wide as a living room. Palomar remained the world's largest telescope for the next quarter century.

Still, Carnegie astronomers had long wanted an observing station in the Southern Hemisphere so they could access the many mysterious objects hidden over California's horizon. In the 1980s they reluctantly abandoned their custody of Mount Wilson, from which stars now appeared dimmer than those in nearby Hollywood, thanks to runaway development and a yearly 10 percent growth in the number of streetlights.[3] Instead, they looked to another site—a mountain in the Andes that the Carnegie Institution had acquired in 1969, when the peso exchange rate was so low that you could buy a genuine sugar-filled Coke for three cents. Named Las Campanas—"the bells"—the observatory soon became the institution's main facility, on which it constructed two of the world's largest telescopes.

The twin 250-inch giants were completed in 2002. Collectively called the Magellan Telescopes, they boast reflectors that are remarkable for their picture-window half-degree fields of view that can take in the entire moon in a single photograph. The exquisite images come courtesy of unique computer-driven pistons that deform the mirrors twice a minute to maintain their perfect parabolic shapes. Equally renowned is the site's rock-steady image quality, unsurpassed in the world. It might as well be outer space.[4]

Such a top-tier research center allows no casual visitors, but I knew I could use my astronomy press credentials to spend a few nights up there. It's the ideal place from which to probe the fastest speeds in the universe. After I chatted with Wendy Freedman, whom I called at her office in Pasadena, arrangements were made. I headed for South America.

The seemingly endless flight to Santiago was followed by good fortune. Las Campanas's director was in town, and thus I met Dr. Miguel Roth for dinner at an outdoor table in one of the many beautiful neighborhoods of that fascinating city. Director for seventeen years, Roth is obviously very proud of the facility: "We're up at eighty-five hundred feet, and it's really dark. The site is incomparable. We get three hundred clear nights a year. The Atacama Desert stretches all around. The nearest corner store is one hundred miles away."

Two days later, after an unnerving flight that skimmed past the Andes' jagged snowy peaks, and after glancing around the cabin to appraise the potential tastiness of my fellow passengers, I arrived in the lovely seaside resort town of La Serena, home of the Las Campanas headquarters. That year, the staff was spending much of its time seeking supernovas, whose reliable "standard candle" brightnesses help determine exact galactic distances, which in turn lets scientists understand how the universe's expansion changes with time. This, then, is the Carnegie Observatories' main current quest—to decipher the fate of the universe!

And thus we reach the meat of this matter. It is nothing less than the greatest conundrum in all of science, and it revolves around speed. Happily, it can be simply stated. The Hubble constant—the speed at which galaxies rush away from us—mysteriously changed six billion years ago, when the universe was half its current age. Galaxy clusters started increasing their flyaway speeds, as if they all had rocket engines that suddenly ignited. The cause is often called dark energy, but that term is no more than a label affixed to an enigma first uncovered in 1998. As Wendy Freedman said with a sigh, "It's very difficult to explain. It's a perplexing mystery."

With a quick revision reminiscent of the hastily airbrushed deletions in Soviet encyclopedias, cosmologists abruptly rewrote their "basics of the universe" handbook. Three-quarters of the cosmos was now exclusively reserved for some kind of weird anti-gravity entity whose existence was utterly unsuspected a year earlier. Probing its powerful effects became a sudden, urgent focus for astronomers. I suspected that this quest, above all, was what occupied those who awaited me atop that Chilean mountain.

The next day I left La Serena in a rented car heading north-bound on a sparsely traveled section of the Pan-American High-way. The road immediately entered the southern edge of the vast Atacama Desert, the driest place on earth. Two desolate hours later I turned onto a relentlessly climbing dirt trail, passing wild burros and an animal called a viscacha, which looks like a cross between a squirrel and a rabbit and which seemed like a hallucination. The broad, bone-dry summit of Las Campanas was dotted with white domes. The high altitude and low humidity created a cloudless azure sky.

I had arrived at noon. Perfect timing. This is when everyone has just awakened. All freshly showered and hungry, they file into the spacious dining hall like some religious cult. Their language resembles English, but the dialect is peppered with esoteric astro-physical terms.

Astronomers Dan Kelson and Barry Madore, Wendy Freed-man's husband, sat with me. It was a precious opportunity, and I wasted no time plunging into profound issues involving cosmic velocity and what it might imply for the future of the universe. "I'm here for the ride," Madore said with a modest laugh. "I'm not here for ultimate answers."

But later, under the stars, he turned serious. "We're living with uncertainty with the universe's expansion," he said after I'd joined him in an enormous dome whose humming computer fans and drive motors formed the sound track to our conversation. The uncertainty involved not just *when* the cosmos went from slowing down to speeding up but also whether the acceleration would con-tinue or even ultimately reverse itself. Still, I thought, if a little uncertainty was the worst he had to deal with, he shouldn't com-plain. Wasn't it enough that humans dared scratch the surface of these fastest of all velocities? Entire cities of suns that speed fifty thousand miles farther away from us *each second?*

I was happy that such a major facility devoted its resources — a legacy of generous endowments that started with Andrew Carnegie's fortune — to such a seemingly intractable quest, and I said so.

When asked to compare Las Campanas to publicly funded instutions, Madore said, "This Magellan telescope costs forty thousand dollars a night to operate. But we can still be playful and innovative and take some risks. That's one big difference. The national observatories, like Kitt Peak — they're all risk-averse. Here it's a thrill a day."

Night had brought a nourishing darkness to the Andes and the unseen black desert below us. The Milky Way — whose name had probably not taxed the astronomy muse — was astoundingly brilliant, with richly mottled detail, as in a pointillist painting. It dominated the Chilean night.

Now, on the catwalk outside one of the 6.5-meter giant telescopes, Miguel Roth joined me, and we gazed up like the Mesoamericans of old, who regarded the Milky Way as the center of all existence.

Miguel had given me carte blanche to roam, so I drove as instructed, with just my fog lights on, along the curvy mountain road that has no guardrails, violating every rule in the driver's ed handbook. I went from one dome to another and visited the researchers in each. At one of the 6.5-meter instruments, I found exactly what I'd been seeking. Here the faint light from distant galaxies, amplified and enhanced one million times by the huge, twenty-foot-wide telescope mirrors, had been accumulating for hours but still had nine hours more to go; astronomers had nothing to do but wait and chat.

My lunch companion, Dan Kelson, was gathering the light from galaxies eight billion light-years away. He noticed my reporters' notebook and started explaining: "This instrument is measuring four thousand galaxies at once. It's an all-you-can-eat type of data collection."

He was used to this endless cycle of data harvesting followed by intense analysis. A brilliant and articulate thirty-eight-year-old from Illinois, he had helped pioneer the new technique of cutting thousands of precisely positioned slits into a metal plate so that a particular group of galaxies can be analyzed simultaneously. If there was anything noteworthy about one city of suns floating in a field of thousands, like a single sunflower in a van Gogh painting, it would immediately pop out to be flagged for further study.

"When I was seven or eight my grandparents got me a refracting telescope from Sears," he later told me. "I studied the lore of each constellation. I read every astronomy book in my elementary school's library."

He was hooked. Kelson earned his doctorate at the University of California, where he simultaneously met his future wife and pursued his parallel obsession with making ice cream: he consumed hundreds of quarts annually.

But all those kilos of saturated fat didn't slow Kelson's passion. His dissertation research involved many nights on the new telescopes at the Keck Observatory, atop Hawaii's Mauna Kea, as well as analyzing Hubble Space Telescope data.

He was exactly the kind of person to whom Hubble would have wanted to pass the torch—exactly the right man to clarify Hubble's bombshell of an exploding universe. His ability to merge cutting-edge spectral techniques with digital analysis formed the ideal skill set with which to follow the galactic footsteps of the legendary long-gone Carnegie astronomers.

By the first light of dawn, Kelson would be detecting objects rushing away from us at the astounding speed of 112,000 miles per second. That is more than half the speed of light. Kelson had in fact, some years earlier, discovered the farthest and fastest galaxy ever known. And his team did it again in 2013, making headlines around the world.

Some of that night's dimmest smudges might lie at the very edges of the observable universe and be the fastest objects humans can *ever* see. This is the outer boundary of velocity — the motion envelope within which everything else dwells.[5]

And yet, astonishingly, these galaxy clusters aren't really moving at all. Rather, *the space between us and them is inflating*. The galaxies are just sitting inertly, like Scrabble players waiting for a vowel. Each is gravitationally jostled by its companion galaxies, but the truly ultrafast speeds we see are an expanding-space phenomenon.

Of course, one might wonder how, if space is mere emptiness, it can expand on its own. How can nothingness do anything? Even with a mandate to explore all manner of motion, it's still odd to discuss the *animation of nothingness*.

But space is not nothing. There's no such thing as nothing. Turns out space has properties. Virtual particles — subatomic particles that live for evanescently tiny time periods and then vanish — pop in and out of existence. Nothingness has inherent energy, and lots of it. According to current theory, an empty mayonnaise jar containing only vacant space has enough energy to boil away the Pacific Ocean in less than a second.

This so-called vacuum energy, or zero-point energy, pervades the cosmos. Thus seeming nothingness seethes with power. And whatever it is, it grows bigger and bigger.

Square one in our natural-motion board game, therefore, involves not just the very fastest velocities but also the frenzied animation of emptiness.

The most frequent question cosmologists get is: What is the universe expanding *into?*

For many, it's the most perplexing motion-related inquiry, and scientists hear it routinely. However, to ask such a question means you've pictured the universe as an inflating balloon that you're viewing from the outside. In actuality, no such perspective exists. There is no "outside" to the universe, by definition. The conundrum arises because the questioner has set up a nonexistent vantage point.

Instead, one should visualize living within a galaxy cluster and observing all the others. We see them all flying directly away from us. Distances between clusters are growing everywhere. This is the basic truth, and we can all picture it. And whether we deem the galaxies to be moving or the space between there and here to be inflating, the result is the same. The gap between us and distant galaxies is steadily growing.[6]

Moreover, the *rate* of the universe's size increase is itself growing. We're living within an ever more powerful self-perpetuating explosion. Most astronomers think it's caused by that mysterious antigravity force pervading every cosmic nook and cranny, dark energy, which keeps rearing its invisible head. That's probably what started everything blowing outward from the get-go. In a very real sense, the big bang is still banging. This runaway mushrooming of the entire universe is the picture frame that surrounds all other movement.[7]

Our exploding universe, which also contains small regions of contracting, collapsing entities, results from a tug-of-war between phantoms in black robes, in which dark matter does most of the pulling and dark energy does the repelling. The latter is winning the contest. Dark energy first gained the upper hand six billion years ago, even if we only just learned the news around the time we were switching from dial-up to broadband.

If we could someday gain light-speed capability, which physicists assure us is impossible, we *still* couldn't reach the farthest visible galaxies, not even if we traveled forever. Thanks to the acceleration of the expanding universe, by the time we arrived at the galaxies' *present* location, which would require more than thirty billion tedious years in our spaceship, the distance between us would have increased so enormously that *they'd be farther away than ever.* Here is futility beyond even the petty frustrations of Sisyphus.

Indeed, the light of those galaxies we were trying so hard to reach would no longer be visible. The trip would be worse than pointless. Our quarry would simply have vanished without a trace.

Lest we let ourselves feel too crushed by this news, within these dizzying extreme-motion cosmic parameters lie astounding secrets we *have* uncovered. Kelson himself promised to reveal some when his data was complete. But, as I was to learn, the true story of nature's motions and speeds did not unfold without laughable errors, egotistical ambitions, and unspeakable tragedy.

The mistakes and the head-scratching began long ago. The oldest Hindu religious text, the Rigveda, written in Sanskrit around 1500 BCE, pondered how it is that "the waters glide downward to the ocean." By the time Old Testament books were penned, a key point was not motion but its opposite. Psalm 93:1 says, "The world also is established, that it cannot be moved." The universal assumption was of a stationary earth. The sun circling around us while our planet remains motionless seemed beyond dispute, because even an idiot could watch it happen. You could see stuff in the sky moving, and you could feel that we were *not* moving.

The standard wisdom was that, as in everyday life, the fastest-seeming objects must be those closest to us. (A car going down your street changes its angular position faster than a plane in the sky.) To the ancients this meant that the moon must be closer to us than the stars. It daily traverses twenty-six of its own widths as it speeds through the constellations. At the other extreme were the six thousand glowing dots whose patterns never changed; they must lie farthest away. This assigning of distance—the moon nearest and the stars farthest—ultimately proved true. So the ancients managed not to be wrong about *everything*.

By the time of the Greeks, the stars that circled us nightly were assumed to be inlaid into a kind of crystalline sphere—check another box in the "incorrect" column. But with the tools at hand

2,300 years ago—i.e., none—how could anyone *begin* to figure out the truth?

Yet that is exactly what one Greek accomplished. I introduce him proudly, because he is my first hero.

Aristarchus of Samos, born in 310 BCE, pondered these moving entities in the sky and arrived at correct conclusions eighteen centuries ahead of everyone else. A mathematician and astronomer, Aristarchus was the very first person to say that the sun is the center of the solar system. And that Earth orbits around it while also spinning like a top. To his contemporaries it must have seemed nothing short of crazy. And indeed, with fellow Greeks Plato and Aristotle contradicting and even ridiculing him, Aristarchus's insights—based on lunar shadowing and the relative positions of the sun and moon—didn't "take off" until another seventy-two human generations had come and gone. Even Aristarchus's contemporary and fellow Samos native Epicurus—yes, *that* Epicurus, who was fond of life's pleasures—claimed the sun hovered nearby and was just two feet in diameter. *Two feet!* Early evidence, perhaps, that hedonistic ouzo binges are not compatible with math.[8]

Meanwhile, odd celestial events, such as eclipses, along with earthquakes and other scourges, were usually seen as a manifestation of anger from God or the gods. It became our human task to figure out why the deities were so enormously ticked off and to appease them. Moreover, for more than thirty centuries, natural events that either threatened life or were considered capable of doing so—and these included comets, planet conjunctions, eclipses, storms, and epidemics—were regarded as omens. They didn't just happen, they had meaning. Omen interpretation was a popular activity and, for those with the gift of gab, a lucrative business. There was no word in either Greek or Latin for "volcano"—which illustrates how little importance was paid to the physical event as opposed to the presumed underlying cause, divine fury.

Meanwhile, to rational Greeks, the issue of what moves and what doesn't remained secondary to the basic question of why anything should move in the first place. It may have seemed an insoluble problem, but Leucippus, and especially his student Democritus, who was born around 460 BCE, originated and popularized the idea that everything is composed of infinitesimally small moving particles called atoms. Each is colorless and indivisible, they said, and when atoms glom together to form the various objects around us, those objects mobilize as a result of their atoms' motions.

This atomic theory became a popular explanation of nature's animation. The reason the belief lasted only as long as the modern "Elvis is alive" concept—a couple of generations—is that it ultimately collided with the genius of Aristotle.

Aristotle was born on the Greek mainland in 384 BCE. His prolific writings were a mixed bag, although many of his goofs were inherited from his teacher Plato. These mistakes included relatively minor errors, such as his belief that heavy objects fall faster than light ones, and major blunders, such as his insistence that Earth is the stationary center of all motion.

He spent countless pages exploring the causes and nature of motion in his groundbreaking book *The Physics*. In one of its subsections (Book II) Aristotle claims that actions begin because nature "wishes to achieve a goal."

But here's the point. In both Greek atom theory and Aristotelian philosophy, natural motion originates *from within each object*. This is the reverse of our modern thinking. Science now insists that nothing budges unless acted upon by an *outside* force.

Many of Aristotle's ideas provide grist for deep thought to this day. Book IV discusses *time as being a quality of motion,* which, he said, has no independent existence of its own. He also implied that an observer is necessary for time to exist. Both these concepts are very much in line with modern quantum thinking. Few physicists today think that time has any independent reality beyond being a tool of animal perception.

Later in *The Physics,* Aristotle tackles the old "prime mover" enigma by arguing that the universe and its motions are eternal. You don't need an initial instigator to start the ball rolling. Everything moves; it's always moved; it's its nature to move.

In other words, as we gaze at nature's endless animation, we see a pageant that has no need for the cause-and-effect business: every moving entity exhibits the dynamic aliveness of the eternal One. It sounds very much like Hindu Advaita or Buddhist teachings.

Moreover, Aristotle said, matter's energy never diminishes. In this, too, Aristotle is confirmed by modern science. We have accepted since the nineteenth century that the universe's total energy never decreases.

With this mixed assortment of profound and nutty notions, Aristotle is nonetheless best remembered for yet another aspect of his epic treatise on why things move: the elements. He actually borrowed the idea from Empedocles, who was born around 490 BCE in Sicily. Embraced for the next two thousand years, this theory basically states that everything is made of earth, air, water, or fire (or mixtures of them), to which Aristotle added a divine fifth element, ether, found only in the heavens.

Aristotle said that each element *liked to exist in a particular place and would always go there if it could.* This, he said, was the central reason for motion.

A clay pot, for example, is made of earth. This element fundamentally belongs to a realm at the center of the universe (i.e., beneath the ground) and hence desires to return there. So at the slightest provocation a pot falls because *its natural motion is downward.* That would bring it closer to "home."

The element water also wants to go down. Its domain is the sea, which, for the ancients, was the region surrounding the lowest realm—made of earth, clay, and rocks. This is why people, composed of lots of water, easily fall and bruise. Our bodies *want* to fall. But when bathing in the ocean we don't fall or even necessarily sink because our body's watery element is now "home" and at rest in its natural environment.

Fire, on the other hand, is of a mysterious realm high above us, and thus its natural motion is upward. This explains why fire and anything associated with it, such as smoke, readily rise. The element air is another lives-up-high substance, which explains why bubbles in water always head upward.

Hence was born the notion of "place"—everything has its preferred position and tries to go there. Aristotle said that natural place has a *dunamis,* or power to create motion.

It all made sense. It *still* makes sense, even though it's wrong. Aristotle's notions about why things move held sway for eighty generations, until well into the Renaissance. It was still the paradigm for a no less brilliant observer than Leonardo da Vinci, who made frequent allusion to the four elements.

Leonardo's writings make contemporary motion beliefs crystal clear. In particular, he articulately pondered the nature of *force:*

> "It is born in violence and dies in liberty."
> "It drives away in fury whatever opposes its destruction."
> "Force lives always in hostility to whoever controls it."
> "It willingly consumes itself."
> "Always it desires to grow weak and to spend itself."

Judging by these quotations from a 1517 Leonardo manuscript, it's obvious that he saw *force,* another initiator of motion, as an almost sentient presence. It had deliberate objectives. Like Mona Lisa, it schemed and dreamed.

It took another 170 years—until 1687, when Isaac Newton

spelled out his three laws of motion in the *Principia*—for modern concepts of how and why things move to finally appear.

Of course to us, as observers and participants in nature's non-stop action, the real enjoyment lies in simply watching the pageant. And here, I was to learn, even sluggishness brings jaw-dropping surprises.

CHAPTER 2: *Slow as Molasses*

How We Learned to Love Lethargy

I'm ready to go anywhere...
— BOB DYLAN, "MR. TAMBOURINE MAN" (1964)

The human brain has a bias. We are wired to notice abrupt motion.

If we stare blankly out a window, thinking about our taxes, we'll be snapped to attention if the still life is punctured by sudden movement. A rabbit darting from a bush, say. The scene may already be pregnant with countless slow movers—caterpillars, subtle swayings of branches, clouds mutating—but we will be oblivious to all of it. A shame. While we may pay attention to the sudden fast things, Earth's oozing, creeping entities influence our lives far more than the darting bunnies.

Our bias toward speed is at least as old as written language. Though the pace of life in olden times was far more leisurely than it is today, classical and ancient literature showed little interest in "slow." True, everyone knew the sun set 180 degrees away from where it first appeared at dawn. And agricultural societies cared about wheat growing taller. But only the final result mattered. They didn't know or care that corn grows an inch a day, an imperceptible motion twenty times slower than a clock's hour hand.

We are all prisoners of our experience, and human motion was the standard for what we'd call slow or fast. The speediest person who ever lived is alive today: Jamaica's Usain Bolt. He ran the hundred-meter dash in 9.58 seconds in 2009 for a speed of twenty-three miles an hour. This is the very fastest a human has traveled using no more than his own legs. As if to prove it was no mere ephemeral fluke, he virtually equaled that speed when he left all competitors behind during the 2012 Olympics in London.

Of course, nobody can maintain that pace for long. The fastest mile, at three minutes and 43.13 seconds, amounts to sixteen miles per hour. And the best a marathoner has achieved is to average 12.5 miles per hour. We consider animals slow or fast based on the ancient important issue of whether they can catch up to us from behind.[1]

But our exploration at the moment is of the far more prevalent realm of slothfulness. Speaking of which, those three-toed mammals didn't earn their reputations for nothing. A sloth, even when motivated, only walks at 0.07 miles per hour. "Breeze it, buzz it, easy does it" — as Ice sang in *West Side Story;* the most excited sloth would need a long summer day to cover a single mile. Even giant sea turtles lope 25 percent faster.

Perceived speed is a tricky business. We regard something as fast only if it moves its own body length in a short time. For example, a sailfish swims ten of its lengths per second and is thus viewed as very swift. But a Boeing 747 airliner approaching for a landing only manages to traverse *one* of its lengths, 230 feet, in a second. It's visually penalized by its own enormity. From a distance, a descending jumbo jet seems virtually motionless because it takes an entire second to fully shift its position. Yet it actually moves four times faster than the fish.

Now consider bacteria. Half the known varieties have the ability to propel themselves, usually by whipping their flagella — long helical appendages that look like a tail. Are they slow? In one sense, yes. The fastest bacteria can traverse the thickness of a human hair each second. Should we be impressed?

Zoom in, however, and this motion becomes remarkable. First, that bacterium has moved one hundred times its own body length each second. Some manage two hundred body lengths. Relative to their size, they swim twenty times faster than fish. It's equivalent to a human sprinter breaking the sound barrier.

Moreover, the covered distance quickly adds up. Germs can transport themselves one or two feet per hour. That's the speed of a minute hand on a wall clock. No wonder diseases spread.

In our homes, other eerie motion unfolds as well, all the time. Dust in the air, for example, much of which consists of tiny bits of dead skin. Watch a sunbeam cast its rays through a window and your home's omnipresent suspended dust becomes obvious. After all, light rays are invisible in and of themselves. In our homes we see a beam only when it strikes countless slow-drifting particles. In very humid conditions, minuscule water droplets catch the light. But in dry air it's always dust.

A quick glance makes it seem as if the suspended particles aren't going anywhere. They move up or down with the slightest air current. But leave the room alone—at night, for example, when nobody is disturbing anything—and all this dead skin and other detritus settles at the rate of an inch an hour. That's ten times slower than all those scurrying bacteria. Who suspected that our homes are so creepy?

In the visible realm, the standard archetypes for intimate slow mo are our fingernails. And hair.

Fingernails grow a quarter of an inch longer every two months. That's half the rate of hair growth. If we neglected our barber appointments the way Newton and Einstein did, we'd find our hair six inches longer each year.

But nails vary in interesting ways. Our longer fingers grow their nails more speedily. Pinkie nails advance sluggishly. Toenails grow at only one-fourth the rate of fingernails. That is, they grow at that rate unless you like to walk barefoot, which stimulates growth. Fingernails respond to stimulation, too. That's why typists and computer addicts enjoy the fastest-growing nails of anyone. Maybe this explains why so many of us writers like to bite them.

Nails grow faster in summer, faster in males, faster in nonsmokers, and faster in pregnancy. But nails do not grow at all after you're dead. That macabre myth probably started because the skin on dead fingers pulls back, exposing more nail within two days after a person has passed away.

Probably the most dramatic example of slow motion on earth is the earth itself. In caverns, stalactites and stalagmites typically extend at the rate of one inch every five hundred years. By comparison, mountains are downright speedy; they push themselves higher—in the case of the Himalayas, anyway—by a couple of inches a year.[2]

Stalactites are mirrored in this reflective pool in Luray Caverns, located in Virginia's Shenandoah Valley. It typically takes five hundred years for each of these downward-pointing structures to grow one inch.

A 2006 study showed that mountain ranges typically rise to their full height in only about two million years. Mount Everest has grown measurably taller since it was first scaled. Some activities just keep getting harder.

Actually, you yourself are moving even when you're doing the couch-potato thing. All landmasses are shifting, carrying you and your TV toward the west if you live in the United States. You can lie in bed and sing, "California, here I come!" But at half an inch a year, you'd better bring your own trail mix.

This tectonic drift was first discovered by Abraham Ortelius, a well-regarded Flemish mapmaker, in the late sixteenth century. He wrote, "The Americas were torn away from Europe and Africa... by earthquakes and floods" and went on to note that "the vestiges of the rupture reveal themselves if someone brings forward a map of the world and considers carefully the coasts of the three [continents]."

Independently, Alexander von Humboldt, in the mid-nineteenth century, while mapping the eastern coast of South America, wrote that its emerging outline seemed like the adjoining jigsaw-puzzle piece for the western side of Africa. The only logical conclusion was that continents shift. But neither of these men was credited with this astonishing revelation. Nor did any other scientists take the idea and run with it. It wasn't until Alfred Wegener's 1912 theory of continental drift that people started taking it seriously, even if there remained more critics of it than believers for the next half century.

Here was a case where you had an effect — landmass motion — before you had any conceivable cause. Yet it always stared us in the face. What's below Earth's surface? Lava, obviously — what we now call magma. This is a liquid. Suddenly it seemed plausible that continents float on this thick, dense fluid. And if they float, they obviously could shift. The problem was coming up with a mechanism or force that could propel them sideways. Ever try pushing a stalled car? Imagine the torque required to budge an item like Asia. Continents are not pond scum.

That's why the idea of drifting continents was not widely accepted among the top geologists. It was, in fact, ridiculed for decades. No proposed mechanism that seemed truly plausible came forth, at least none in which the math would work. It took until the 1950s and particularly the 1960s before the true reason for landmass motion finally came to light. The cause had been hidden beneath thousands of feet of murky brine.

It was the dramatic but unknown reality of the sea floor spreading apart. Mid-ocean volcanic activity creates widening fissures and forces a growing separation between the floating continents. The greatest fault line, the Mid-Atlantic Ridge, is the primary point of separation for the earth's crust. New techniques of seismology and, finally, GPS tracking sealed the deal.

Nowadays we know of eight separate floating landmasses, each chugging along in various directions. The Hawaiian chain is the fastest moving, as it heads to the northwest at the rate of four inches annually. We can now also easily match geologic features on one continent's edge with those on another's, proving they were connected in the not-so-distant past. For example, eastern South America and western Africa not only share specific unique rock formations but also contain matching fossils and even living animals found nowhere else. Similarly, the Appalachians and Canada's Laurentian Mountains are a perfect continuous match with rock structures in Ireland and Britain. All the evidence proves that the separate continents were once a single supercontinent—the famous Pangaea. It formed three hundred million years ago and started breaking apart one hundred million years later.

Before Pangaea there were long periods of multiple drifting continents separated by several oceans alternating with periods in which single, unbroken supercontinents were surrounded by water, which encircled the entire planet. The monolithic supercontinents that preceded Pangaea have names like Ur, Nena, Columbia, and Rodinia. We humans got to see none of it. Even the Rodinians, 1.1 billion years ago, never strutted around with proud *R*s on their sweatshirts. They were microscopic creatures who lived exclusively in the sea.

So in continental drift we have something continuous and certain that vastly changes the appearance of Earth over tens of millions of years. Here is slow, epic, ceaseless movement—unseen and unfelt. And suspected by not a single pre-Renaissance genius.

In our human obsession with measuring and categorizing things, we find one very obvious end point when it comes to speed: The bottommost terminal. Nothing can travel slower than "stopped." Yet it's surprisingly hard to find anything that exhibits no motion on any level.

If we look closely, even a sleeping sloth stirs. It's breathing, and its atoms jiggle furiously. But it's especially cool to note that the colder something is, the slower its atoms move, so true motionlessness means reaching a state of infinite cold.

At the chilliest place on earth (the Antarctic, where a frosty negative 129 degrees Fahrenheit was registered in 1983) there's still plenty of atomic motion. Atoms stop moving only at 459.67 degrees Fahrenheit below zero. That's *absolute zero*. It was first recognized by the brilliant if cantankerous Lord Kelvin in the mid-nineteenth century; his posthumous reward was the increasingly utilized Kelvin temperature scale, which places its zero at that momentous point (rather than at water's freezing point, as Anders Celsius did, or at the temperature of an icy brine slush, which is where Daniel Fahrenheit chose to position his scale's starting position).

Until the mid-1960s, astronomers thought that if thermometers were positioned far from any stars, they would register absolute zero throughout the universe. Now we know that the heat of the big bang produces a five-degree warmth that fills nearly every cosmic crevice. It's usually expressed as 2.73 degrees on the Kelvin scale. (And the universe keeps getting colder all the time, chilled by its expansion like a discharging aerosol can of whipped cream: it was twice as warm eight billion years ago.)

The universe's coldest known place, its ultimate Minnesota, is right here on earth, in research laboratories where temperatures less than a billionth of a degree above absolute zero were first created in 1995. This technological deep freeze yields an Alice in Wonderland of bizarre conditions. When atomic motion stops, matter loses all resistance to electrical current, creating superconductivity. Strange magnetic properties also arise (the Meissner effect), making magnets levitate like magician's assistants. Then there's superfluidity, in which liquid helium defies gravity and flows up the sides of its container, escaping like some resourceful mouse by simply scampering up and out. Finally, a new state of matter materializes as any substance approaches absolute zero. Neither solid, liquid, gas, or plasma, it's called the Bose-Einstein condensate. Shoot light into it and the photons of light themselves come to a virtual halt.[3]

But any exploration of nature's slowest entities would be incomplete without an examination of the single substance most associated with lethargy. We're talking about molasses. Slow as molasses.

Being scrupulously scientific, our quest was to find any actual measurement and qualification of molasses's exact viscosity (gooeyness). Unearthing this information wasn't easy. One paper from 2004, in the *Journal of Food Engineering*, had this soporific abstract:

> The rheological properties of molasses with or without added ethanol were studied using a rotational viscometer at several temperatures (45–60°C), different amounts of added ethanol in molasses-ethanol mixture per 100 g of molasses

(1–5%) and rotational speed ranging from 4.8 to 60 rpm. Flow behaviour index of less than one confirmed pseudo-plasticity ($n = 0.756 – 0.970$)...

And on it went until the punch line was reached at last:

The suitability of the models relating the apparent viscosity were judged by using various statistical parameters such as the mean percentage error, the mean bias error, the root mean square error, the modeling efficiency and chi-square (χ^2).

Okay, so how viscous or slow is molasses? And what is it, anyway?

Molasses actually has three forms, all of which result from sugar refining. Essentially you crush sugarcane and then boil the juice, extract and dry the sucrose, and the leftover liquid is molasses. The initial fluid remnants are called first molasses. If you then reboil it to extract more sugar you get second molasses, which has a very slightly bitter taste, and I hope you're taking notes on all this. A third boiling of the syrup creates blackstrap molasses, a term coined around 1920. Since most of the sugar from the original sugarcane liquid has now been removed, the blackstrap molasses is a low-calorie product, thanks to its skimpy remaining glucose content. The happy news is that it contains the good stuff never removed in the processing, including several vitamins and major amounts of minerals such as iron and magnesium. But we don't really care about all this. What matters is how slow it is.

Viscosity is a liquid's or gas's thickness. Its degree of internal friction. The less viscous a fluid is, the greater its ease of movement. A viscous fluid will not just "run" more slowly, it will also exhibit a dramatically smaller splash when poured.

Naturally, we often use water's viscosity as a comparison. If water is rated at just below 1 on the scientific viscosity rating scale, then blood officially rates a 3.4. So blood really *is* thicker than water.

Sulfuric acid has a viscosity of 24. Did you know that that scary acid is so syrupy? Thinnish winter-use motor oil with an SAE 10 rating has a viscosity of 65. In contrast, the thick motor oil used in hot regions, SAE 40, has a very high viscosity of 319. This is stereotypical guy stuff.

Enough fooling around. Here are the truly slow-flowing fluids:

VISCOSITY OF COMMON FLUIDS

Olive oil	81
Honey	2,000–10,000
Molasses	5,000–10,000
Ketchup	50,000–100,000
Molten glass	10,000–1 million
Peanut butter	250,000

So forget molasses. "Slow as peanut butter" would better designate a refusal to flow. However, most people might not regard a jar's worth of Skippy as a liquid and would thus disqualify it from the motion contest. When bored kids can use a spoon to create firm peaks in a substance, it's hard to deem it a fluid.

Molasses did have its fifteen minutes of fame. That's when it dramatically defied its reputation for sluggishness. The most spectacular molasses event in world history happened in Boston on an unusually mild January day in 1919, just after noon. At forty-three degrees Fahrenheit, it was ten degrees warmer than normal for that time of year. That's when an enormous, poorly welded six-story-high cylindrical tank suddenly ruptured on Commercial Street, near North End Park. Two and a half million gallons of molasses burst out. People fled, and it would have been a funny sight had it not been for the tragic loss of twenty-one lives as men, women, teens, and several horses were engulfed and drowned in the sticky tidal wave.

Above the disaster scene, a packed elevated train had just passed, its incredulous passengers witnessing the collapsing tank and the approaching black wall of ooze. The viscous fluid broke the steel supports of the elevated train structure. When the trestle snapped, the tracks collapsed nearly to the ground—but the train had already sufficiently advanced to remain safely aloft a few hundred yards farther along.

The expression *slow as molasses* was already firmly in the public lexicon in 1919, and even the estimated thirty-five-mile-per-hour speed of the four-story wave of fluid, which—only because it had been piled so high—caught up to everyone trying to flee from it, failed to erase molasses's clichéd reputation as the epitome of lethargy. The idiom remains.

Of course, when we think of peril instigated by a sluggish liquid, molasses doesn't usually spring to mind first. Lava does.

And when it comes to cataclysm, no event before or since has remained as rooted in the collective consciousness as the total destruction of Pompeii and Herculaneum.

Let's place ourselves in the mind-set of ancient Rome during the year when Titus became emperor, 79 CE. This was a tumultuous time, because before assuming the throne, during the brief reign of his father, Titus had simultaneously led the successful war against the Jews and destroyed Jerusalem and had a scandalous affair with the Jewish queen Berenice. Talk about having it both ways. Then, just two months later, Vesuvius blackened the skies. Few at the time failed to link that cataclysm with the gods' apparent opinion of the seemingly dubious character of the new emperor. And his troubles were just beginning.

Of course, we may wonder why even today anyone in his or her right mind would choose to live in, say, Naples, just five miles from an active volcano whose tendency is toward explosive, Plinian-type eruptions. Let alone purchase real estate on its very slopes — the location of the unfortunate Pompeii and Herculaneum.

The 79 CE eruption of Monte Vesuvio (or, in Latin, Mons Vesuvius) changed the course of the Sarno River and raised the ocean beach, plunging property values in several ways. Afterward, Pompeii was no longer either on the river or adjacent to the coast.

Science lets us look back to the mountain's genesis. Modern core samples drilled 6,600 feet into its flanks and dated using potassium-argon techniques show that Vesuvius was born from the Codola Plinian eruption just twenty-five thousand years ago, even if the entire region had known general volcanic activity for about a half million years.

The mountain was then built up in a series of lava flows, with some smaller explosive eruptions interspersed between them. The game perilously changed about nineteen thousand years ago, when Vesuvius's regular eruptions became more explosive, or Plinian, events.

Before Pompeii, the Avellino eruption, about 3,800 years ago, destroyed several Bronze Age settlements. Archaeologists studying this in 2001 uncovered thousands of preserved footprints of people who were all apparently trying to flee northward, toward the Apennine Mountains, abandoning a village that, like Pompeii, was to be entombed beneath countless tons of ash and pumice. Fast-flowing pyroclastic surges were deposited ten miles away, where modern Naples has, unfortunately, now been built.

Any resident of Pompeii and Herculaneum who could read the classics would have had ample reason for concern about that location. A Plinian eruption a mere three centuries earlier, in 217 BCE, had produced earthquakes throughout Italy, and Plutarch wrote of the sky being on fire near Naples.

But by 79 CE, the lower slopes of the mountain were covered with gardens and vineyards nourished by the rich volcanic soil,

41

with its high levels of nitrogen, phosphorus, potassium, and iron. It was a thriving, popular place. On flat areas near the summit, guarded by steep cliffs, Spartacus's rebel army had their encampment just a few years earlier, in 73 BCE.

Then came the ascension of Titus to the throne and, a mere eight weeks later, the cataclysm that killed at least ten thousand people. We don't have to guess what happened. In a letter to the Roman historian Tacitus, Pliny the Younger gave a gripping first-hand account:

> [A] black and dreadful cloud, broken with rapid, zigzag flashes, revealed behind it variously shaped masses of flame.... Soon afterwards, the cloud began to descend, and cover the sea. It had already surrounded and concealed the island of Capreae and the promontory of Misenum.... The ashes now began to fall upon us, though in no great quantity. I looked back; a dense, dark mist seemed to be following us, spreading itself over the country like a cloud. "Let us turn out of the high-road," I said, "while we can still see, for fear that, should we fall in the road, we should be pressed to death in the dark, by the crowds that are following us."

Pliny continued:

> We had scarcely sat down when night came upon us, not such as we have when the sky is cloudy, or when there is no moon, but that of a room when it is shut up, and all the lights put out. You might hear the shrieks of women, the screams of children, and the shouts of men; some calling for their children, others for their parents, others for their husbands, and seeking to recognize each other by the voices that replied...[4]

* * *

One hundred and fifty miles (by road) to the north in Rome, the clouds from the eruption as well as frantic word of it arrived almost immediately. Titus responded quickly.

Although the Roman imperial administration hardly has the reputation today of having been a compassionate FEMA-type agency, many emperors actually did react generously to natural disasters. Facing a great calamity, Titus appointed two ex-consuls to organize an impressive relief effort and donated large sums of money from the imperial treasury to aid the volcano's victims. Moreover, he visited the buried cities soon after the eruption and again in 80 CE. It should have been enough to assure his popularity, but he couldn't keep up with the plagues.

Calamities kept arriving. The Vesuvius disaster was followed the very next year by a major fire in Rome. Then, in a pattern radiating outward from the fire-ravaged area, no doubt following the paths of the escaping rats, came a deadly outbreak of bubonic plague. Even omen interpreters who liked Titus could find no way to get him off the hook. It seemed the empire would continue to be stricken as long as he sat on the throne. He then made it easy on everyone. He contracted a sudden fever the next year and died at the age of forty-two. Following Titus's brief, frenzied reign, the mountain remained tranquil for more than a full human life span.

The peace was, of course, temporary. In December of 1631, following more than three centuries of total inactivity, a Vesuvius eruption caused widespread damage, and the focus on the mountain was renewed, albeit through the lens of modern scientific inquiry. This inspired many scholarly papers, especially from the numerous academies of Naples, as Renaissance scientists tried to learn how Earth metamorphoses from cold and stable to fluid and fiery. It was only then that science began to pin down the facts of the destructive events of August 24 and 25, 79 CE; something approaching absolute knowledge was not achieved until the 1990s.

Today we know that the eruption was a two-act tragedy. First came the Plinian phase, in which hot matter explosively blew upward in a tall column only to spread out and fall like hailstones. This initial mushroom cloud of material resembled a modern-day nuclear explosion. It began at midday on August 24 and quickly rose to a height of 66,000 feet. During the next eighteen hours it produced a dark, ominous rain of ash and pumice — mostly south of Vesuvius, thanks to the direction of that day's winds. All told, the falling pumice stones, each about a half inch wide, buried Pompeii to a depth of eight feet and yet posed little initial danger to human life.

The eruption's first few hours unfolded in slow motion. Many people, huddling in homes that were being rapidly buried by the fallout, held out hope of survival while keeping their fingers crossed that their roofs might bear the weight even as structures around them began to cave in. Modern estimates show that roofs of that era would start to fail when sixteen inches of pumice settled on them, which would produce a load of fifty-one pounds per square foot. Even the rare, super-well-built timber roof would collapse under the full weight of the pumice layer, at which point it would be asked to bear the impossible load of 476 pounds per square foot. This far exceeds current building codes for concrete warehouses. Thus no structure in Pompeii could have survived the initial, Plinian phase of the eruption, and, we must expect, residents would have been forced to flee, thus making it likely that they did not even witness the next stage, in which another four feet of heavier gray pumice was added, like the frosting on a sadistic cake.

By the next morning, August 25, eighteen thousand Pompeii residents were running for their lives. We know this because only some two thousand bodies were found in the buried ruins of the collapsed roofs and floors. Now came the tragedy's superlethal Peléan phase.

Unfortunately for the residents who hadn't already been killed, this deadly second act featured pyroclastic flows and surges—avalanches of scorching gas and dust plummeting down the mountain at sixty miles per hour, hugging the ground.[5]

These gases—superheated to 750 degrees and intermingled with nearly red-hot dust particles—produced most of the fatalities. The gas-and-dust mixture seared lungs. A single breath was lethal. There was no possible defense. The intense heat, exceeding that of an oven, turned much of the area's organic material into carbon. Many of the victims were found with the tops of their heads missing because their brains boiled and exploded in their skulls.

Thus slow-moving lava—which later killed more than one hundred people in 1906, when a Vesuvius eruption produced the most lava ever recorded, and which continues to destroy property on Hawaii, thanks to Kilauea—was *not* the main culprit in the 79 CE Pompeii event.

But "slow-moving" doesn't automatically mean "benign," as bacteria show. Indeed, motions too leisurely to perceive are today causing ongoing worry for millions. And one such motion unfolds in a place few have ever visited—a location that might not even exist tomorrow.

CHAPTER 3: *Runaway Poles*

They're Really Shifting — Are We Toast?

At the still point of the turning world.
—T. S. ELIOT, "BURNT NORTON" (1935)

Four inches a minute.

You'd think that would be too sluggish to bother anyone. And yet no natural motion generates more paranoia than POLES ARE SHIFTING headlines, a rare marriage of New Age scare literature and mainstream science. Not content to worry about high blood pressure, some people fear we may be on the brink of a global cataclysm whose fault lies not even in the stars but in Mother Earth.

Here are true, measurable changes with a bread-crumb trail that all can follow. It's not a question of if or when. The poles *are* shifting. And they've never moved faster than they do today. Is this concerning?

It's fun and easy to become the neighborhood shifting-poles expert. There are just a few things to learn. One of the facts, a basic one, entered our cerebral cortices in fourth-grade earth science. This is the knowledge that our planet has *two sets of poles*. They bear no resemblance to each other. Both are always in motion, yet their effects are utterly different.

The poles that are moving wildly and with unprecedented speed and that generate all the "what does it mean?" anxiety—those are the magnetic poles. But let's begin with their competitors, the poles of rotation, a.k.a. the geographic poles.

46

These poles are where our planet's spin axis intersects the surface. They're where meridians of longitude converge into a pinprick. The addresses are latitude 90° north and latitude 90° south. These are the only places that have no longitude.[1] The northern address is where Santa lives. When you stand at the North Pole, every direction, any step you take, is south. No other direction has any meaning. You can turn off your car's GPS.

At the poles, and only at the poles, you're not carried along by our planet's spin.

If you live on the equator, you whiz eastward with the earth's rotation (which is separate from our planet's devilishly fast 66,600-mile-per-hour orbital motion) at 1,038 miles per hour. This barely changes as you travel—*at first*. Go five degrees, or 350 miles, north, and the spin speed lessens by a negligible four miles per hour. But the next 350 miles brings it down by twelve miles per hour. By the time you reach Boulder or Brooklyn you're moving at just 795 miles per hour, and a further 350-mile jump slows you another sixty miles per hour, taking you below the speed of sound for the first time.

(Where on earth would you be rotating at *exactly* the speed of sound? Woodstock, New York. The laid-back hippie place, still into music. Who says irony isn't everywhere?)[2]

The planet's rapid slowdown at higher latitudes soon gets out of hand. Fairbanks spins at just 422 miles per hour. It's zero at the pole. You'd just stand there nearly motionless, like an idiot, pivoting too slowly for anyone to notice, facing the opposite way twelve hours later.

Since 90 percent of all humans live in the Northern Hemisphere, let's surrender to our boreal bias, apologize to our Aussie, Kiwi, South African, and South American friends, and focus on the North Pole for the sake of brevity. You reach it by heading due north from anywhere. Reindeer do not live there. No one does. It lies in the Arctic Ocean, which used to be frozen all the time but nowadays is open water during the summer, when the kids are home.

Becoming the first person to reach that spot was once the most prestigious thing you could possibly do. A century or so ago it would make you an instant Neil Armstrong. The problem was that, with no communication with the outside world and no humans within a thousand miles, you and whomever you could talk into joining the party would be utterly on your own. Even if you could somehow contact the Royal Geographical Society's help line and reach a real person in a cubicle, you can imagine how thrilled they'd be with your problem.

"Hmm. Remember Beardley? Seems he's trapped in the ice two thousand miles northwest of Iceland. Wants to know if we can send someone to fetch him. Yes, right away."

Henry Hudson, itching for more adventures after discovering the river that Sully Sullenberger eventually landed in, managed to come within seven hundred miles of the North Pole in 1607. This was an amazing accomplishment for the time. (In the spring of 1611, on an ensuing polar expedition to find the fabled northwest passage to China, his crew mutinied and put Hudson, his teenage son, and a few others off in an open boat into what we now call Hudson Bay. They were never seen again.)

A few Russian and British explorers managed to inch a few miles closer during the next two centuries, but not by much. The race heated up, if that's even the right word, in the late nineteenth century, when the American James Booth Lockwood and his sledding party got farther north than anyone before him. He died in April of 1884, at the age of thirty-one, during a miserable three-year expedition, two months before a rescue party arrived. He reached latitude 83°24'30", just 450 miles short of the goal.

Two years later, the rhyming Norwegian explorers Nansen and Johansen reached latitude 86°14' north on skis and dogsled from the ship *Fram*, which had been caught and held as if by a vise in drifting ice in the Arctic Basin. They were forced to stop a mere two hundred miles short.

The North Pole was finally attained on April 6, 1909, by the American Robert Peary, who succeeded by dogsled. The South Pole was conquered just two years later, in December of 1911, by the expedition of the Norwegian Roald Amundsen. A month later, the pathetic Brit Robert Scott arrived; his infamous quest not only failed to get him there first but cost the lives of the entire party, thanks to the brutal bad luck of hitting the coldest weather in a decade.

These explorers reached poles that are not nailed in place. Despite their reputation as the steadier of the two types of poles (it's the magnetic poles that are moving like crazy), these geographic poles do shift back and forth. It isn't much; there's a deviation of only about forty feet per year from the average position. The maximum movement is barely over one hundred feet. However, because astronomical and geodetic measurements, not to mention deeds and survey maps, are based on the latitude and longitude system—those horizontal and vertical lines on maps and charts, pinpointing every pond, home, and used-car lot on the planet—and because polar shifts throw these numerical positions slightly out of whack, the variations gather a lot of attention and are constantly monitored.

To obtain accurate information about polar motion, the International Latitude Service was established in 1899; it was renamed the International Polar Motion Service in 1961. Nowadays the ongoing quest is handled by the International Earth Rotation Service. You want to know what our planet is up to? They're the ones who will tell you. The service's outposts are dedicated to making continual observations of latitude changes (and rotation hiccups and lots of other arcane stuff). They update the precise coordinates of the geographic poles all the time. This is a serious group, whose members grab the phone and say, "Are you sitting down? You won't believe this!" if the poles unexpectedly shift a foot, as they do after some major events, such as tsunamis.

The geographic, or physical, poles only "jump" following an especially violent earthquake that redistributes the mass on our spinning globe. More usually, their motion consists of just two smoothly changing components. There's a circular shift called the Chandler wobble, which completes a cycle every 433 days, or 1.2 years, and there's an annual movement that goes back and forth in seemingly random directions. The annual component varies from year to year, though it's always less than fifty feet.

Why does Earth do this? Theories have come and gone for centuries. There's no hard evidence for anything. One favorite notion was that Earth's oval orbit makes the solar gravity stronger in January than in July. Nowadays we think it's due to seasonal redistributions of ice and air masses.

And the Chandler wobble? Named for the American Seth Chandler, who discovered this planetary gyration in 1891, it was finally explained 110 years later, in 2001. Richard Gross of the Jet Propulsion Laboratory, using computer simulations, produced a persuasive analysis showing that most of the 433-day wobble comes from changing pressures on the ocean floors caused by variations in temperature and salinity. Salt shifts the poles! The rest of Chandler wobble results from atmospheric fluctuations.

The geographic poles (a.k.a. the physical poles, or the rotational poles) have never shifted significantly, at least not since the moon's creation four billion years ago. And they never will. You could walk across each pole's greatest positional change in twenty seconds.

Obviously, when people speak fearfully of the poles shifting, they can't mean the geographic poles. In reality, most folks don't even know *which* poles worry them. It's been too many years since earth science. If you ask them, "Which poles do you mean—the geographic or the magnetic?" you'll likely get a blank stare.

* * *

A major sudden shift in the geographic poles would create global destruction, but it's never happened and is physically impossible. So enough about them.

Time, then, for the poles that don't just wander in circles a few dozen feet, like my mother-in-law searching for her parked car. We come now to the *animated* poles.

These are the places where compasses point.

Sloshing liquid iron three thousand miles below the surface, moving around the solid iron ball that sits at Earth's center like an olive pit, generates a rather wimpy magnetic field. (Our planet's field averages about 0.5 gauss on the magnetism scale. By comparison, a strong refrigerator magnet is 100 gauss.) Even when a magnetic sliver is balanced on a needle so that it can swivel at the slightest provocation, it only feebly aligns north. There's not much oomph in our planet's magnetism — unlike the magnetism on, say, Jupiter.

Here's a bit of arcana known only by your college physics professor: Earth's *south* magnetic pole is the one that's located in the far north. (A south magnetic pole is one in which the field lines go down, toward Earth's center.) But to keep our citizens in their normal, happy, unconfused state we call this one the North Magnetic Pole, and I have no problem with that. It's located in the north, and that's good enough to merit the label.

As kids, when we'd sprinkle iron filings over paper with a magnet placed beneath it, we'd see the distinctive curving shapes of the magnet's field. And, similarly, Earth's magnetic field lines run horizontally over most of the planet and then dive straight down at the, ahem, North Pole. So a magnetic pole is simply where field lines are aligned up and down. But you don't have to bother trying to visualize magnetism angling into the ground. There's a simpler way to find that "vertical-field-lines" place: compasses point there.

Does it matter where the North Magnetic Pole is located? Not really. Auroras form a glowing green ring around the place, and they're beautiful. People in Fairbanks would be sorry to see it migrate too far. But other than that, if it shifts position it affects nothing and no one.

Good thing, too. The North Magnetic Pole is in constant motion, though it has dwelled in Canadian territory since at least before the days of Galileo and Shakespeare. It currently sits five hundred miles from the pole of rotation. That's closer together than the two competing poles—geographic and magnetic—have been at least since English sounded like anything intelligible to modern ears.

The North Magnetic Pole actually drifted south during the seventeenth and eighteenth centuries until it sat at latitude 69° north, barely in the Arctic at all. Then it started going north. Its location at Ellesmere Island was first discovered by explorer James Clark Ross in 1831. A century ago its northerly motion started accelerating and inexplicably grew from five miles to thirty-seven miles a year. Nowadays the North Magnetic Pole is just west of Ellesmere, the planet's tenth-largest island, where only 140 winter-sports-loving people live, most of them in the Canadian military. They like to boast that they're the most northerly group of humans in the world.

During the last century the North Magnetic Pole has berserkly sped a whopping 650 miles almost due north, so that it's now passing latitude 84° north. It moves twenty-two feet an hour!

In the ocean, it recently crossed Ellesmere's two-hundred-mile territorial limit, which means the North Magnetic Pole no longer belongs to the Canadians. It was one of their claims to fame, and they're not happy about this development. First they have a bad maple syrup year, and now this.

If the pole keeps going in the present direction, it will shoot clear across the Arctic Ocean and down the other side into Siberia by the time today's teenagers start bleaching off their tattoos, around midcentury.

The Buddha preached equanimity—that we should stop having opinions about absolutely everything. Well? Should anyone care about this? Does it matter whether these poles stay put or wildly scamper to a new spot? Here's why some people do indeed care.

Two or three times each million years, on average, Earth's entire magnetic field reverses its polarity. In other words, if you were alive a million years ago, and facing what we call north with a compass in your hand, the needle would point to the south. No matter that you and I can't even sense Earth's magnetism in the first place. Nor can most animals.[3]

The idea of a "pole flip" *sounds* dramatic and worrisome, but actually the science behind it is very cool.

We've only known about magnetic reversals since 1959. They weren't easy to detect at first, because there hasn't been a single one for the past 780,000 years. Turns out, however, that when lava containing ferrimagnetic minerals solidifies, its iron bits line up with Earth's prevailing magnetic field. This happens as soon as the lava cools below its *Curie temperature* of 1,414 degrees Fahrenheit, or 768 degrees Celsius. So we can read these rocks as though they were a novel.

Researchers dug deeper and excitedly turned the pages. Here was a reversal, and here, and here, until they'd unearthed 184 polarity reversals in the last eighty-three million years.

These pole flips were some sort of strange new motion cycle. And you *know* how we humans love patterns and trying to see if they sync up with others. If it were Christmas, these rhythms would be fabulous puzzle toys tied up with ribbons.

But as we unwrapped each one, it became increasingly clear that the poles reverse randomly. There is no rhyme or reason. No pattern. Using radiometric dating, we found that the North and South Poles changed position every 450,000 years on average. But sometimes they'd flip rapidly. A little less than two million years ago there were five reversals in a mere million years. Another time there were seventeen in a three-million-year period. The rock records even revealed a case of two flips in fifty thousand years.

Each geomagnetic period is called a *chron*. In between chrons, there is a transition phase: it takes between ten thousand and one hundred thousand years for the new polarity to establish itself. Contrary to today's paranoid news reports, a magnetic pole reversal was never something that unfolded while you were having a latte. If a pole reversal had begun during the Last Glacial Maximum, when New York's Central Park lay submerged beneath a mile of ice, the reversal would still not be fully established today.

Researchers also found two or three *superchrons,* a period when the same magnetic alignment lasted more than ten million years. The Cretaceous Superchron endured for forty million years. An earlier one had lasted fifty million years.[4] Their causes are anyone's guess. Who can fathom what exactly is going on 1,800 miles beneath the surface, where the liquid outer core begins? Analysis of seismic echoes provides a crude picture of Earth's interior layers, and we can only hope that future refinements will let us understand the hows and whens of magnetic-field production and reversals. One thing is certain: epic patterns of uncountable tons of moving liquid iron are responsible, and they have their own natural animation. The reversals certainly do not correlate with major meteor impacts, sea-level fluctuations, or any other sporadic global events we can find. Nor do they match up with our planet's orbital shifts or changing axial tilts, which, being recurring and predictable, would have resulted in regular rather than random polarity shifts.

The only real concern has been the state of our magnetosphere during the process. What if our field temporarily vanished? Isn't that our protection against cosmic radiation? Would that roast earthly life and afflict us with runaway mutations and cancers?

This remains the basis for the current near hysteria in some overcaffeinated circles.

The scientific answer: there's no problem. If they were truly harmful, reversal periods would match up with times of mass extinctions. They don't. The fossil record shows that pole flips never affect the biosphere. Those weren't even periods of sudden appearances of new life-forms. Evolution wasn't being prodded in those interchron periods.

Recently there's been more reassuring news. It seems our magnetic field does not vanish during those centuries when a reversal is trying to establish itself. Rather, many new magnetic poles chaotically come and go, with our field altering its appearance but remaining more or less intact.

Anyway, analysis shows that even without a magnetosphere, our atmosphere blocks most incoming radiation. We'd lose only a frosting of protection. It's like a professional carpet cleaning — nice but not strictly necessary.

It's been 780,000 years since the last pole reversal. A long time. Pole flip or no, these have been good years for us mammals. We're slightly overdue for the next reversal as far as long-term averages go, but we stand nowhere near the fifty-million-year duration record. And when the flipover process starts, it could last for a thousand centuries.

Has it indeed started? One strange fact is that our global magnetic field has become 10 percent weaker since 1850. And the poles are indeed changing position very quickly. But contrary to the beliefs of those who see meaning — usually grim meaning — in the myriad physical events unfolding around us, no one really knows if these are signs of anything. Maybe Earth's magnetic strength always fluctuates up and down and this 10 percent business is perfectly normal. And maybe the poles sometimes shift position rapidly, sometimes slowly. There's no way to know if ours is an unusual period.

Even when it does happen, you'd never be able to tell that you were living in an interchron time of magnetic pole reversal without using measuring equipment. Maybe some pigeons would fly confusedly, but that would be it.

Colonel Norm Couturier, commanding officer of Canadian Forces Northern Area, who is the man charged with protecting Canada's Arctic sovereignty, was interviewed about the runaway North Magnetic Pole for the *Edmonton Journal* in 2005.

"It's a force of nature that we're not equipped to deal with," he said jokingly.

Admitting that it would be sad to lose the pole, Couturier pointed out that it also has a bright side: with the pole gone from Canada, Canadians have less responsibility for the ill-prepared adventurers who go on half-crazed skiing adventures to reach the magnetic pole.

"It will probably mean now that we'll have to stage less rescue missions," he said. "When it was over in Canadian territory, every year we would have to go and assist somebody or recover somebody that was trying to get there. Now that it's in international waters, a little bit of the pressure is off us."

There you have it. The runaway pole is making *some* people happy.

CHAPTER 4: *The Man Who Only Loved Sand*

And the Curious Phenomena of the Atacama Desert

You throw the sand against the wind,
And the wind blows it back again.
—WILLIAM BLAKE, "MOCK ON, MOCK ON,
VOLTAIRE, ROUSSEAU" (CA. 1800–3)

The Atacama—a region of utter stillness but occasional strange phenomena—is the driest place on earth. Like its host country, Chile, it tirelessly runs from north to south, occupying a vast expanse centered about twenty degrees south of the equator, the latitude that's home to nearly all major deserts no matter what continent they're on. Its main geographic oddity is that it's narrow: the Atacama begins without preamble at the western base of the Andes and stops suddenly after a mere sixty miles, at the cold Pacific.

Having left the Chilean Andes, I had no choice but to drive into the Atacama. I had exercised questionable judgment, however, by impulsively choosing an untraveled sandy trail that headed north and west. It was designated by the thinnest possible line on the map, snaking along for about seventy miles before it reached a seaside fishing village. My gas tank was nearly full, I had a bottle of water, and, well, what else do you need?

Yet after a mere hour of solitary driving among nothing but dunes and small rocks, not encountering a single car passing me going the other way, the initial elation of adventure was replaced by a vague and uncomfortable feeling.

No one was there. The sun was fierce. There was certainly no cell-phone service. What if the car broke down? No one knew my plans. When would another car travel on this dry dirt trail? Would the next vehicle appear any time this month? This year? I glanced at the single plastic liter of water on the seat next to me. It suddenly occurred to me that I was an idiot.

Turn around or keep going? I figured I was roughly at the half-way point. It made no difference now. Anyway, never turn back. At this point, I had no idea that my odyssey would become bound up with that of a legendary British brigadier named Ralph Bagnold.

Suddenly, without preamble, a yellow dust devil appeared perhaps forty yards in front of me, and I slammed on the brakes, creating a competing dust cloud. It was a dead ringer for a tornado, a miniature version. I got out and had to crane my neck to see how very high it towered into the blue cloudless sky. And now it was joined by a twin to my right. Swirling crazily, they both moved ahead at maybe walking speed and showed no sign of dissipating. Each was perhaps six feet thick. In the unchanging sameness of the desert, where everything else was utterly motionless and even the wind was perfectly calm, this sudden lively animation was startling. The fierce whirlwinds were not just surreal. To tell the truth, they were downright spooky.

Unlike tornadoes, dust devils develop from the ground up. They favor dry places, such as deserts, and do not form from clouds. Indeed, like the pair I was now observing, they usually materialize beneath calm, cloudless skies.[1]

I knew that they could reach above the tallest skyscrapers, but these towered perhaps three hundred feet. Thirty stories.

In the dry, very thin air of Mars, the sudden materialization of dust devils marching across the chocolate-orange soil seems like the work of spirits. In fact, these dust devils are referred to in Arabic as *jinni,* which means "demon" and which was the origin of our word *genie.* They abruptly give that lifeless red planet the brash hint that, yes, the hand of nature still stirs even there, where Earth is a mere dot in the sky.

Those bizarre "spirits" may even be benevolent. On March 12, 2005, technicians monitoring the Mars rover *Spirit* found that a fortunate encounter with a dust devil had blown off the thick dust on its solar panels, which had choked off much of the power supply. Now, suddenly, electricity generation dramatically increased. Expanded science projects were joyously scheduled. Previously, another rover, *Opportunity,* had also had its solar panels mysteriously cleaned of accumulated dust, and a dust devil was likewise assumed to have been the cause.

I got the sudden urge to step into one. Would it be dangerous? How fast were those winds, exactly?[2]

I'd heard that dust devils sometimes throw jackrabbits into the air. But the only truly scary story I'd ever encountered was that of three children who sat in an inflatable playhouse just outside El Paso, Texas, in 2010. The trio were taken into the sky, playhouse and all, carried over a fence and three houses, and then deposited on the ground without serious injury.

The impulse was irresistible. It was my investigative duty as a science journalist, I rationalized. I trotted clumsily across the sands toward the nearest dust devil, my feet sinking with each step, but the whirlwind moved away like a mischievous *jinni.* It kept eluding me, and then the farther one dissipated abruptly, as if in a dream.

When I finally turned back toward where I thought the car and dirt road were, there was no trace of either. They must be hidden in a depression, I thought. With the lone dust devil snaking away, the silence of the harsh, sunlit desert was overwhelming.

I stood, mesmerized by the isolation. I felt quarantined from the human race.

Anyone who has been to a desert knows its hypnotic appeal. In 2006 I had gone to the Sahara to meet the moon's shadow, but that total solar eclipse wound up as merely the initial enticement for me; the desert's magic only grew. And much earlier, as a twenty-two-year-old wandering the world with a backpack, I spent a couple of weeks in the great vast desert of southeastern Iran, between Kerman and Zahedan, where the nightly skies are as inky black and star-filled as I imagine they are on the far side of the moon.[3] I had also loved the Thar Desert, with its herds of wild camels and friendly people, in the Rajasthan wastelands of western India. Every desert is unique. This one, the Atacama, is special in several ways.

For starters, it's the driest of them all. In some sections no measurable rain has fallen for the past five years. There is thus not the slightest trace of even scrub vegetation. The cold, rich South Pacific, with its famous Humboldt Current, laps at the desert's beaches, where penguin colonies nest in its protected bays, while the forbidding peaks of the Andes abruptly define its eastern edge. These mountains are the culprits that manufacture the aridity. The prevailing easterly winds are forced to rise, cool, and dump their moisture on southern Bolivia and northern Argentina. At night, nearly continuous lightning over the Andes hovers above the unseen border between the two countries. When the air descends from the Andes it is bone dry.

Lack of rainfall and vegetation are the calling cards of most deserts, but they also hold one other feature in common: the classic stage set of blue sky and fierce sun. With no trees to cast shadows, the Atacama offers no relief.

Standing amid a 360-degree panorama of stark, sandy, sunlit isolation, I realized I'd overlooked the most central "action figure" of all, the one whose natural motion rules everything else. The sun.

The author appears lost in the desert. On all the world's ergs, sand moves in a precise, mathematical way.

For most of us, it is sometimes factored into our plans: Should we cancel our beach trip if it's cloudy? But it's relatively rare for the sun to modify our behavior in modern times. We mostly ignore it. Even science nerds are only vaguely aware of its various cycles and quirks.

But here in the desert there is nothing else. The sun calls all the shots. And if you're stuck without a means of exit, it ultimately decides whether you live or die.

Its most basic animation is the day-night rhythm. This remains steady throughout our lives but is much less uniform over the lifetime of our world. When the first dinosaurs walked through the Meadowlands of New Jersey — then part of the supercontinent Pangaea — the year had four hundred days.

Not impressed? Then look back further, to when life first appeared. Earth spun *much* faster then. The environment was truly

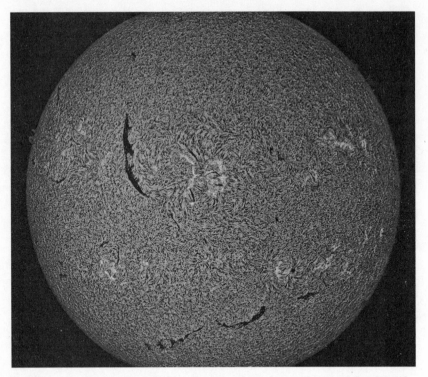

As the sun spins once a month, its surface pulses up and down like that of a subwoofer. *(Matt Francis)*

alien, unrecognizable as the precursor to today's world. The air had no oxygen. The sun was 30 percent dimmer and daily crossed from horizon to horizon in five hours. *It visibly moved.* Shadows perceptibly shifted, as they do in time-lapse nature photography.

The moon's tidal tug creates an oceanic bulge beneath it and a second bulge on the exact opposite side of Earth. These bulges travel around as our planet spins, exerting a bit of torque as countless tons of seawater smash into coastlines to deliver the "High tide!" news to bathers and gulls. Continuously lengthening our days by slowing our spin, the moon's tides make the sun move ever more sluggishly across the sky.

We're reminded of this every year or two when scientists announce the insertion of a "leap second" into the final minute of

June or December. Television stations give this job to their meteo-rologists, who explain that extra seconds are needed because our planet is winding down and will ultimately make each rotation, each of our days, forty days long in the far future.

But if you're a serious card-carrying geek you've surely stopped in your tracks, grabbed your calculator, and said to everyone within earshot, "Wait a minute! They add a second every year or two? Earth can't be slowing that quickly. It just can't!" You go *tap-tap-tap* on the instrument's keys and realize that if our planet's day was really growing a second longer every couple of years, we'd have come to a frozen halt billions of years ago. Something doesn't add up. Something about the rotation of our world simply doesn't make sense.

Because the media always get this wrong, here's the real scoop. The answer involves beauty. Poetry, even. After all, a watch set to the right time is a device synchronized with Earth's rotation. It lets Orion and the Dog Star march to the beat of the timepieces on our wrists or, more likely these days, the überprecise digital time on our smartphones, whose signals are periodically synchronized with atomic clocks even if we don't care about such accuracy.

In the 1950s an important decision was made, an agreement between every nation on our whirling planet. It was, simply, that Earth's spin rather than vibrating quartz crystals or any other time-keeping method would dictate the time. This meant we needed two parallel monitoring systems kept in sync with each other. One is our planet's spin, constantly scrutinized by an agency in France called, not surprisingly, the International Earth Rotation Service.

The other system requires the careful daily marking off of 86,400 seconds, each precisely defined. These official ticktocks are counted by forcing the nucleus of the cesium 133 atom to maintain a particular spin direction, which it does only when bathed in 9,192,631,770 microwave pulses per second. Any other frequency changes the cesium. So an atomic clock is simply a vacuum chamber where a fountain of gaseous cesium atoms are bathed in microwaves and the state of the cesium is continuously monitored. That's the story. A servomechanism slightly varies the microwave frequency if required. An official second is thus 9,192,631,770 microwaves, just what's needed to maintain cesium 133 in a fixed condition. That exact number of microwave pulses *is* the definition of a second.

The official second remains constant. Earth, alas, does not. Along with spin irregularities not fully understood, observations of the stars show that our planet's day becomes one seven hundredth of a second longer after each century has passed.

This may seem too trifling to matter at all. Compared to the day you were born, the day you start receiving Social Security checks is one thousandth of a second longer. Sure, this adds up, but it's way too little to require meddling with clocks every year or two. So again, why those leap seconds?

Here's the explanation that, guaranteed, nobody on your block knows.

When the current system was set in place in the 1950s, astronomers had been using earth-rotation data collected over the previous three hundred years. The official length of a day was codified in 1900. But during those centuries of observation, a day's length slowly grew. Careful analysis now shows that a day was exactly 86,400 seconds long in 1820. Before that each day was shorter. Since then it's been longer.

We generally labor under the illusion that 86,400 seconds make up a day. But this hasn't been true for nearly two hundred years. A modern day is 86,400.002 seconds long. So we messed up. When the current system was put into place a half century ago, we *could have* then defined each second a little differently by adding a couple hundred more of those microwave beats to each official second.

Who would care? Then our clocks would almost never need leap seconds. But we didn't. So every year or two now, the little daily error accumulates enough so that we must take care of the accrued discrepancy.

To sum up, the real problem is not that Earth is slowing, which happens too gradually to matter much. It's that each of our current days is longer than a day was in 1820, upon which our timekeeping system is, bewilderingly, based.

Because we foolishly designed the "second" around the 1820 data, we now need to compensate for the difference between a day now and a day when James Monroe was president. That means adding a second every five hundred days or so. It's a "patch" to keep Earth-spin time and atomic-seconds time in agreement.[4]

As the Earth slows, the sun moves more leisurely across the sky. Its gradual slowdown is of course unnoticed in human lifetimes. Instead the dominant rhythm that affects us is its position in the sky. The sun is low and feeble during winter and high and fierce in summer. The daily light-darkness ratio—winter's short days and long nights—is also critical. Other than that, most folks are oblivious to the sun's motion. How many people even realize that throughout the Northern Hemisphere, in the United States, Europe, China, and so on, the sun *always moves to the right?* Meaning the sun rises diagonally upward to the right, then moves directly rightward at midday, and sets by slinking rightward into the western horizon.

Equatorial residents view something different. There the sun rises straight up until it gets overhead. Then, through the afternoon, it drops straight down like a lead ball. Because of this, it quickly buries itself below the horizon after sunset. Twilight in the tropics is always short. In the Southern Hemisphere, the sun moves *leftward* during the day. It's a quick way of knowing where you are in case you're ever shanghaied and wake up on another continent.

* * *

Can you handle one more solar oddity? Over the course of a year, day and night are *not* balanced. Thanks to our atmosphere, which bends light, the sun seems to sit on the horizon when it's actually already set. At that point we see a ghost, a solar phantom. This air trickery, *refraction,* grants most locations seven minutes of extra daily sunlight. It's why days and nights are *not* equal at the equinoxes: sun dominates.

This undeserved sunshine adds up. We enjoy *forty extra hours of sunlight annually.* The year is not even close to a fifty-fifty day-night mix.

On top of that, as we all know, sunset is never followed by sudden blackness. On the moon, yes, but not here. Refraction delivers its enchanting gift of twilight. Its brightest portion bestows yet another hour of useful light split between dawn and dusk.

The brightest afterglow is called *civil twilight.* Although it sounds vague, the term *twilight* is precisely, legally defined, dictated by the sun's unseen motion below the horizon. In the evening it's the interval between sunset and the time when the sun has sunk six degrees, or a dozen sun widths. Civil twilight lasts about a half hour in most places. At its conclusion, according to many municipal ordinances, streetlights must be on.[5]

But the bottom-line sun motion is its speed as it crosses the sky. Most people don't know about angles or degrees, so let's simply use the sun's own width as a measuring tool. Think about all the sunsets you've watched. How long does it take the sun to move a distance equivalent to its own diameter? Or ponder the moon instead, which moves at the same visible speed. The answer:

Crossing the sky, the sun traverses its own width in exactly two minutes.

During a sunset, because the sun slides into the horizon at an angle, the interval from first contact to complete disappearance is about three minutes. *This is right on the borderline of perceptible motion.* The sun appears to move at the same speed as the minute hand of a kitchen clock when viewed from a few feet away.

Our final desert-motion phenom is its most renowned specialty: the mirage. As we all know, mirages are common on hot surfaces, such as a highway on a summer afternoon. The culprit is the changing speed of light. Despite its reputation as a constant, light travels more slowly through cool air. But the hot air above a summer road or broiling sand lets light move faster right there, closer to its vacuum speed, and this change bends, or refracts, images hitting it. The result is a mirror effect. The air reflects the sky, perfectly mimicking a puddle of water.

But finding *any* movement was an impossible job when I was in the desert. Nothing budged once those dust devils died. The absence of flowing water, moving clouds, circling birds, buzzing insects, or rustling leaves makes the desert visually frozen. A still photograph. Its landscape offers the antithesis of animation.

But later there came a few hot afternoon gusts. Bits of sand blew momentarily. The still life came alive. Clearly the dunes migrate over time. And when it comes to shifting sands, only one person is associated with their vagaries. British brigadier Ralph Alger Bagnold.

He was the archetypical English stiff-upper-lip, military-cum-Renaissance man. Bagnold was born in 1896, son of a derring-do colonel in the Royal Engineers who gloriously participated in the 1884–85 rescue expedition that attempted to free Major General Charles George Gordon from Khartoum. His sister was Enid Bagnold, who wrote the bestselling 1935 novel *National Velvet*.

Armed with this odd genetic pedigree, Bagnold attended Malvern College, joined the Royal Engineers, as his dad did, and received medals for serving in the miserable World War I French trenches for three years. After the war Bagnold studied engineering and earned a master of arts degree at Cambridge University. He returned to active duty in 1921 and then got swept into his lifelong calling. He served in Cairo and the Thar wastelands of northwestern India, and at both places he spent every spare minute exploring the desert.

Bagnold described his extensive excursions in his book *Libyan Sands: Travel in a Dead World* (1935). He developed a special type of compass that would not go awry around the iron ore often buried in arid regions. It was he who discovered that one really could drive a car across the Sahara as long as you let most of the air out of the tires and kept punching the gas pedal when the sands got deep. You got the feeling this was knowledge gained the hard way.

Although a third of the world's deserts are covered with sand, there has been very little research into these *ergs*, as sand-covered desert areas are, oddly, called — probably because it's hard to travel there or even reach many of them, and at that point it's very slow going to make much physical progress. Bagnold changed that with his still-definitive book, published in 1941, *The Physics of Blown Sand and Desert Dunes,* which is every bit as tedious as it sounds. After the first two or three chapters I discovered that it is not a page-turner, despite the Amazon five-star rating that lured me to purchase it. But no one to this day has improved on its revelations. Bagnold used wind-tunnel experiments to predict sand movement and confirmed these expectations with extensive observations in the Libyan desert.

Basically, sand is characterized by its size rather than its composition. Bagnold defines sand as any particle between 0.02 millimeter and 1.0 millimeter in diameter, although later experts generously expanded the upper range by more than 50 percent, to 1.6 millimeters — a fifteenth of an inch. Size matters, because sand is defined as consisting of grains small enough to be moved by the wind but too heavy to remain in suspension in the air, as dust and silt do. Particles too heavy to be blown by wind are classified as pebbles or gravel. If it's smaller than a thousandth of a millimeter, a particle essentially remains suspended in the atmosphere and scarcely falls at all. But then it's called smoke or dust, not sand. It's not rocket science.

Although sand can be composed of nearly anything, most of it is quartz, essentially because quartz is common and, Bagnold explained, "resistant to both mechanical and chemical breakup into smaller sizes."

Wind, needless to say, is responsible for sand piling up into dunes and also for abrading and rounding each grain. (Sand beneath rivers and seas is another story, of course, because there water is the erosive force.) Because sand is two thousand times heavier than air, it doesn't blow easily. It's not house dust. No action at all is observed when the wind is less than ten miles per hour, which was my initial experience in the Atacama. But then, when the wind blows between ten and twenty miles per hour, lots of activity unfolds all at once.

The wind moves sand in two ways. The main method is called saltation, which is the picking up of grains. In this method the weight of the grains quickly brings them back down a short distance away. Carefully watching the process is like observing the quick hops of millions of kangaroos. The other transport method is called creep, in which wind rolls or bounces grains. In this method the grains typically move forward at about half the speed of the wind. Both courses of transport are in-your-face obvious if you spend time in a sandy desert on a breezy day. There's really no other action to observe.

You'd also think—wandering among the endless dunes—there'd be no possible sound but the wind. That's usually true. But on rare occasions the desert sings. Says Bagnold at the very end of his extensive sand-physics study:

We now pass from the squeaks made by small quantities of beach sand when trodden underfoot, to the great sound which in some remote places startles the silence of the desert.

Native tales have woven it into fantasy...sometimes it is said to come upwards from bells still tolling underground in a sand-engulfed monastery; or maybe it is merely the anger of the jinn! But the legends...are hardly more astonishing than the thing itself.

Bagnold then shares his personal experience: "I have heard it in southwestern Egypt 300 miles from the nearest habitation. On two occasions it happened on a still night, suddenly—a vibrant booming so loud that I had to shout to be heard by my companion."

The security and serenity of the dunes, Bagnold's life's obsession, had been suddenly replaced by spookiness. At the far end of the earth this scrupulous man of science was enveloped by irrational mystery. He knew that sounds are always caused by

Dark horizontal markings on rock structures at Utah's Lake Powell are boundary layers that once existed below the sea. Such long-term alterations in our planet's appearance, unknown to classical thinkers, unfold too slowly for human perception.

motion. But how on earth can dry sand create an earsplitting detonation or the equally bewildering "singing"?[6]

Ultimately Bagnold admitted, seventy-five years ago, that these
desert cacophonies are a mystery. The acoustic puzzle has persisted
through the years despite being the subject of two recent TV specials. Bagnold did, however, notice that the bizarre booming or
singing, which sometimes continued for more than five minutes,
"always came from the lower part of a sand avalanche."

After completing his epic study, Bagnold founded and was the
first commander of the British Army's Long Range Desert Group
during World War II. He wrote scientifically useful papers through
the 1980s before leaving Earth for the great cosmic desert in 1990
at the age of ninety-four.

My special time in the Atacama duly noted, I trudged back to the
now-baking car and reached the fishing village in another hour.
There I met upbeat people who spoke so slowly that even I could
understand their Spanish. I hired a fisherman to take me out to see
penguin colonies in a boat that looked only marginally seaworthy
in the large swells, and with the engine off we sat quietly while several dolphins came to the starboard side, where we listened to them
breathe. Glorious. But this was enough leisurely slow-mo time.
Within a few days I made my way back to Santiago and a flight to
the place where our world turns at its fastest rate.

I had a specific goal in mind: the unique events found only at
the equator. What awaited me instead was a surprise.

CHAPTER 5: *Down the Drain*

Weird Goings-On at the Equator, and the Frenchmen Who Died Young

By heaven, man, we are turned round and round in this world...
— HERMAN MELVILLE, *MOBY-DICK* (1851)

In Quito, perched at nearly ten thousand feet above sea level, the air is so thin that a visitor can scarcely walk two of its hilly blocks before stopping to catch his breath. I was here, very simply, because the capital of Ecuador is the world's only city that sits right smack on the equator—where our planet's spin hurls every pedestrian around the earth's axis faster than it does anywhere else—1,038 miles per hour.[1]

Supposedly the equator also offers unique opportunities to witness Earth's strange effects on moving water. Because human bodies and brains are mostly H_2O, I wanted to see firsthand this relationship between our most intimate companions, the whirling world and swirling water. I'd heard that the Ecuadoran government had built a major museum precisely on the equator and that there were daily demonstrations.

The equator is more than merely the fastest-spinning place on earth, the place where the moon and stars whiz fastest through the sky. Thanks to the centrifugal force that makes Earth a bit like a carnival ride, people at the equator are partially lifted off our world, like those on the periphery of a high-speed carousel. A beefy man

72

weighs a pound less in Quito than he does in Fairbanks, making it a potentially lucrative place for a weight-loss clinic guaranteeing instant results.

Moreover, because of our planet's oval shape—the midriff bulge that makes the earth's diameter at the poles twenty-six miles less than its diameter at the equator—the middle is also where you stand closest to the moon and sun. Those romantic songs about a big tropical moon? They're true—although the difference in size is just 1 percent. I wondered how many such science tidbits would be on display.

When I stepped out of the taxi (I *loved* saying, "Take me to the equator!") I was nearly knocked backwards by the earsplitting sound of a band playing amplified salsa music, demonstrating the region's well-known distrust of silence. The huge complex of plazas,

Outside Quito, Ecuador, the location of the equator is marked by an enormous five-story monument. It's where our planet rotates at its very fastest. But the government built it in the wrong place.

wide granite steps, small gift shops, and open spaces where I stood was called Mitad del Mundo—the middle of the world. At its center, a multistory stone obelisk dominated the scene. A line set into the ground radiated from this monument and extended in two opposite directions for hundreds of feet. Here it was—the equator!

Tourists, mostly from other parts of South America, straddled this line so that they could have their pictures taken with one foot in the Northern Hemisphere and the other in the Southern. With the deafening happy music and bright sunshine and colorful clothing and constant animated laughter in the thin air, this was a destination not of egghead subtlety for geography geeks but of carnival fun.

Except it isn't really on the equator.

Long ago, in the days before GPS precision, the government built the monument in the wrong place. No brochure actually says so, of course; you have to learn it in a whisper from tour guides, who seem secretly thrilled to spill the beans. It didn't appear as if anyone cared.

I quickly learned that the actual equator was up the road, a quarter mile north, and it was there that one could find the moving-water exhibits. Leaving this fancy official government complex with its grand stone structures and busy souvenir stalls, I hiked up the narrow highway until I reached a sign boasting that the real McCoy lay up a dusty dirt road. An arrow on the sign pointed the way. Panting in the thin air while jumping across potholes, I eventually came to a private museum, complete with its own painted equator line. The first lecturer I bumped into said that, yes, modern measurements prove that *this* is the true equator. I checked my handheld GPS and wasn't at all sure she was right.

So far I'd learned only one thing: our planet has competing equators. Each gets many visitors. And I soon gleaned from my GPS and had it confirmed by an official that the true, honest-to-goodness equator is in neither place but rather lies another hundred yards farther north, on vacant grassland. If you're looking for a business opportunity, buy this plot of land, have it paved, and then paint a third equator line. The crowds seem big enough to support lots of them.

The museum hosted nonstop demonstrations, most of which were ridiculous, including a woman in jeans standing at a folding table whose sole job was to balance an egg on a tray and then bilingually claim it can happen only at the equator. I finally reached my mecca, the exhibit that indeed drew twelve-person crowds every fifteen minutes—the supposed proof that water swirls down drains in opposite directions in each hemisphere. Impromptu variations of this demonstration are also held for tourists in numerous African villages. It has become the equatorial de rigueur "thing to do," just as looking for the green flash has changed sunset gazing from the beautifully purposeless activity it used to be into a scientific endeavor.[2]

An attractive young woman leaned over a small, beat-up metal basin on legs and pulled the plug. The crowd watched the water spiral clockwise down the drain and into a big plastic bucket below. Then she and an assistant dragged the basin ten feet across the dubious equator line, the legs scraping cacophonously on the concrete floor while everyone winced. I wondered why they didn't simply purchase a basin with wheels, since they obviously did this day in and day out. The two smiling Ecuadorans poured in the water again, the woman pulled the plug, and sure enough, the water spiraled down the opposite way. The crowd enthusiastically lapped this up (figuratively, of course) with appreciative murmurs. I had to admit it was pretty dramatic and persuasive.

After the group had moved on, and with the next group approaching, I collared the lecturer and quietly said, "May I do this demonstration myself?" Her enormous eyes met mine and showed a hint of alarm. She then held up a "wait just a moment" finger and scurried off to get the director.

Astride the equator, purportedly marked by a line of painted tile on the ground at left, an Ecuadoran woman demonstrates the way water swirls down the drain in a basin. It swirls in one direction on one side of the line and, after she drags the basin across it, in the opposite direction on the other side of the line.

In mere seconds, a twinkly, paunchy, middle-aged man materialized and offered his hand while I introduced myself in my best halting Spanish as a science writer. I might as well have identified myself as a clown, since his reaction was unrestrained laughter. I quickly learned that he was one of those rare fortunate people who find everything in the world amusing.

"Sure, you can perform the demonstration," he said with a chuckle, but then he lowered his voice a bit, glancing at the next approaching group. "Just make sure to pour the water the correct way. Do it from the right when on the far side of the equator line"—and here he gestured as if emptying a pail sideways—"and then pour it in from the other direction when the basin's on this side of the line. That's the only way to make the water go down the way we want it to."

In other words, they're faking the whole thing.

"But this exhibit is a hoax!" I protested. Hearing this, the director laughed so heartily I suddenly wished I could keep him in my life forever. I think if he had somehow asked, I would have given him my daughter's hand in marriage.

"Well, maybe it is!" he said with a giggle, "but we only claim that it's a *demonstration*. If we don't do it this way, it won't work. And the tourists love it." He glanced at a sign near the basin. "How else can we teach them about the Coriolis effect?"

The museum's sign indeed explained that water goes down drains differently in each hemisphere as a result of what it called the *Effecto Coriolis,* which, the sign said, influences many other things, too. (It's certainly influenced livelihoods in many enterprising African villages, where residents perform variants of the same bogus presentation.)

This whole business probably started in 1651, when the Italian scientist Giovanni Battista Riccioli published his book *Almagestum Novum,* which said that a cannonball's trajectory should, strangely, curve to the right because of the spinning of the earth. This was actually a perilous statement, since Galileo had been hauled before the Inquisition just eighteen years earlier and forced to swear that Earth doesn't move at all.

The freedom to speak openly about a spinning Earth had long been established by the time Gaspard-Gustave de Coriolis was born in 1792 in Paris, a few months before Louis XVI was guillotined. A science whiz kid, he placed second in the entrance examination for the prestigious École Polytechnique, became an engineer, and in his young years, despite chronic poor health, made major contributions to various scientific fields that involve motion, such as friction and hydraulics and water wheels.

By the time he reached the age of forty his brilliance was well known to members of the Académie des Sciences, thanks to his groundbreaking treatises on mechanics and motion. Coriolis coined and established the terms *kinetic energy* and *work*—in the physics sense—which are still universally used today. What would he think of next? He managed to surprise and delight the Académie in 1835 with a novel, perspicacious analysis of the math and physics of the game of billiards. (So *that's* how the introverted Gaspard spends his spare time when his wife is out buying chaussures!) It was in that same year that he published the paper that would ultimately gain him sufficient fame for his name to be uttered daily by countless twenty-first-century equatorial tourists. Yet no one today recalls the paper's title, since it seemed deliberately created as a cure for insomnia: "Sur les équations du mouvement relatif des systèmes de corps" ("On the Equations of Relative Motion of a System of Bodies"). In the second of its three sections, Coriolis spoke about the way moving objects reliably curve, though he never mentioned Earth's spin or our atmosphere.

Scientists soon realized that Coriolis had explained perfectly why Caribbean hurricanes always rotate counterclockwise, why artillery shells veer away from their targets, and indeed why a perfectly balanced car speeding along on a level highway pulls, annoyingly, to the right. (An unknown number of wheel-alignment customers no doubt shell out millions of unnecessary dollars annually after being fooled by this effect.) In the early twentieth century, meteorologists started using the term *Coriolis force* to describe the vagaries of large-scale wind and storm systems.

And yet today, the Coriolis effect is routinely misunderstood. Flushing a toilet does *not* result in water spiraling down in any particular direction that corresponds to one's location. The effect does, however, produce such oddities as robbing players of home runs, though Albert Pujols has not yet held up a clenched fist and yelled, "Damn you, Gaspard!" In ballparks in which the batter faces north or south, as they do in Dodger Stadium or Wrigley Field, a ball hit by a bat curves an inch to the right, so that it occasionally goes foul when absent Coriolis it would have stayed fair.

Unfortunately, in the great tradition of making physics as tedious as possible, most Coriolis explanations resort to discussions of inertia, reference frames, angular velocity, and something called Rossby numbers. A pity, since the Coriolis effect is actually easy to understand. Imagine two children riding on opposite sides of a merry-go-round, throwing a ball back and forth. If this carousel is spinning the same way Earth does—counterclockwise, as seen from above—then the child who throws the ball will observe it curve sharply to the right. Aiming it so that it can be caught by her friend would require no small amount of compensation.

Most of the early Greeks imagined that if Earth rotated, a person jumping up would come down in a different spot. But in reality, all objects partake of local motion. Say you live in Miami, where the ground and everything on it zips eastward at 933 miles per hour. Places north of you rotate more slowly; those south of you move faster. Now imagine you've purchased a potato cannon, which uses flammable gas or compressed air to hurl a spud tremendous distances. A half mile would be optimistic, but let's say you've built a Big Bertha model that can throw a good Idaho a full degree of latitude, or sixty-nine miles, and you fire it toward the north. The ground a mere sixty-nine miles north of Miami moves eight miles per hour slower. The potato doesn't know this, so while it's in flight it's also still going rightward, or eastward, at its initial Miami rotation speed. Meanwhile the ground beneath it moves more and more slowly as the flight progresses. Result: the starchy missile goes straight, but anyone on the ground sees it curve to the right.

Say you turn around and aim toward Key West's boisterous Duval Street, to the south. The ground a mere one degree, or sixty-nine miles, south of Miami moves eight miles per hour *faster* than Miami does. So the south-heading spud flies over terrain that's going more quickly than it is. It's being left behind, and as a result it again appears to curve to the right, as witnessed by anyone on the ground.

So flying ballistic objects curve rightward whether they're fired north or south (in our Northern Hemisphere, that is). Only those shot due east or west go straight. This is the Coriolis effect. If the potato hurtles at sixty-nine miles per hour and magically remains flying through the air for an hour, it will land fully eight miles to the right of where it's aimed.

Barring a silo explosion, potatoes aren't generally flying around us. But clouds and air masses are. Imagine a low-pressure storm, like a hurricane. Air tries to rush into it from surrounding higher-pressure regions. As it does so, it flies over ground that is rotating at a different rate than it is. The result is a right-turning tendency. Bingo: a circular storm spinning counterclockwise.[3]

That's why hurricanes never form within 350 miles of the equator. There, not enough Coriolis deflection exists, since Earth's rotation speed is rather uniform in the tropics. There, moving air steers a more or less straight path.[4]

The Coriolis force also explains why everyday winds in most of the United States blow from the west. Air rises up because of equatorial heat and then heads toward the North Pole. Doing so, it deflects to the right. Voilà: our prevailing westerlies.

Now consider your toilet, a form of meditation strongly sanctioned by the likes of Kohler and American Standard. Naturally, the waters on opposite sides of the bowl partake of Earth's rotation. The water on the south side of the toilet moves faster than the water on the north side if you live in North America, Europe, or Asia. Shouldn't this give the water a push so that when you hit the lever it spirals down the drain counterclockwise?

Let's do the math. Turns out the *difference* in Earth's spin speed between one side of a twelve-inch bowl and the other is the same as the speed of the hour hand on a kitchen clock. *The hour hand.* Basically stationary. It's not zero, but it's obviously not going to give ten pounds of liquid any kind of shove. Instead, the direction of swirl depends solely on the direction of the incoming water, dictated by all those little holes concealed beneath the lip of the porcelain. In a basin or tub, the drain-swirl direction is determined by the basin's or tub's levelness.

Gaspard-Gustave de Coriolis never got involved in any of this. He didn't even *have* flush toilets. In fact, though he became a distinguished professor of math, physics, and mechanics and ultimately director of studies at the prestigious École Polytechnique, he never lived to see his name widely attached to the effect he discovered. As in the case of Dorian Gray, his handsome, clean-shaven appearance was deceiving, for his body suffered perennial poor health. He found his energy rapidly diminishing in the spring of 1843 and died late that summer, at the age of fifty-one.

We've seen that although our planet's motion is not manifested in the water swirling in toilet bowls, it is indeed manifested by the west wind blowing daily on our cheeks. But is there any way we can be sure, right here in this room, that we live on a spinning ball? This was an issue that obsessed yet another Frenchman, Léon Foucault, born in Paris in 1819.

Foucault made the first-ever accurate determination of the speed of light, using a polygonal mirror that spun eight hundred times a second. He and another Frenchman were also the first to photograph the sun, in 1845. Back in those days of daguerreotypes, very long exposures were required even for such a brilliant object, and Foucault used a clock drive, a geared mechanism hung below a telescope, to track the sun as it crossed the sky. It was while using this common astronomical device that he noticed that its suspended governing weight, which swung like a pendulum, seemed to gradually change its orientation. With a sudden shock, Foucault realized — a *sacre bleu* moment — that it wasn't the pendulum's path that was changing, it was the ground beneath the telescope. The pendulum was actually maintaining a near-constant plane of swing relative to the universe.

The son of a publisher, Foucault was an innately gifted teacher and popularizer. He wasted little time constructing an enormous pendulum fashioned of a massive sixty-two-pound iron ball suspended from a wire more than twenty stories high. He then set it swinging in the Panthéon in Paris. (Who *does* things like that today?) A sharp metal point welded onto the bottom of the ball scratched a line in the sand he had spread on the floor. Crowds watched as the line shifted, proving that the earth was rotating under the apparatus. Here was the first irrefutable demonstration of our spinning world that could be performed within a single room.

Not only was this pendulum endlessly duplicated and immensely popular throughout the late nineteenth century, it remains so today—even in our age of fast-evolving technology. At a time of severe budget constraints in 2007, when the State University of New York built a new science building at its most prestigious honors college, Geneseo, they scarcely had funds for frills. Yet they nonetheless installed a large brass Foucault pendulum in the lobby, which swings back and forth to greet all visitors.[5]

Despite the fact that he was homeschooled, and despite the fact that he disappointed his family by abandoning his original plan to enter the medical profession (he had a squeamishness about blood that bordered on phobia), Léon Foucault changed the world. It was he who coined the word *gyroscope*, perfected telescope mirrors, and revealed the vagaries of light speed, including the way it slows down under certain conditions. Most of all, he gained global renown for his dramatic pendulum, proof of our whirling planet, for which he received the prestigious Copley Medal of the Royal Society in 1855—that era's equivalent of the Nobel Prize.

Sadly, Léon Foucault fared no better than Coriolis in the longevity department. Mirroring a dramatic swing of one of his pendulums, at the peak of his fame and to the horror of his family and friends, he suffered a sudden and startling physical deterioration that was probably the result of a rapidly progressing form of multiple sclerosis. He died in 1868, at the age of forty-eight.

If you manage to overlook his name near the pendulums oscillating nonstop at science museums throughout our spinning world, you might notice it among the seventy-two names of science luminaries—including Coriolis—inscribed by Gustave Eiffel on the first balcony of his tower.

CHAPTER 6: *Frozen*

The Unhurried Riddles of Snow and Ice

Trust not one night's ice.
—GEORGE HERBERT, *JACULA PRUDENTUM* (1651)

If you look at a map of Alaska and stick a pin right in the center, you'll be close to Fairbanks. In China or India, Fairbanks would be called a town, maybe even a large village. But in this ultra-low-density state, with an average of approximately one person per square mile, Fairbanks officially earns the "city" moniker even though it has a stoic population of merely thirty thousand.

It was late winter, and a short drive on tires that went *thump thump thump* for a while—their flat spots were caused by the rubber freezing during overnight parking—took me away from all traces of Fairbanks. It's easy to leave civilization behind in Alaska. In 2013, taking along a tour group of forty-four adventurers, I'd been heading east for nearly two hours, toward the Yukon. But the sparsely traveled road never makes it there. Or anywhere close. It ends at Chena Hot Springs.

The aurora borealis dances above many places, but nowhere does it appear more often than in Chena, where it is particularly striking against the dark, inky skies. The reason is simple. All auroras are merely small sections of an enormous glowing doughnut that surrounds Earth's magnetic poles. As we've seen, the northern pole lies adjacent to a barren Canadian island in the territory known as Nunavut. Whenever solar emissions get particularly

84

intense, the auroral oval widens and expands southward. Then people in Wisconsin, Pennsylvania, and, in extreme cases, even Florida get to see it from their backyards.

That happens every few years. More usually the aurora borealis forms a steady ring that hovers over the middle of Alaska. Right in the Chena Hot Springs–Fairbanks area. For Fairbanksans, the aurora is more familiar than deer ticks are to northeastern suburbanites.

This was my sixth wintertime trip to this region. I had been the aurora lecturer for an *Astronomy* magazine tour group back in the late 1990s and early 2000s, and now, with solar activity once more on the rise, I was doing it again for a private science company. But this year I had embarked on an additional quest for a specific polar experience: hidden natural *motion* in the white wilderness.

An aurora shimmers over central Alaska in March 2014. Though its motion seems leisurely, this light show is the result of broken atom fragments from the sun striking us at 400 miles per second. *(Anjali Bermain)*

Alaska's vast frozen landscape includes quirky dynamic aspects that lie beyond most people's ken. But this wasteland's animation really starts with the simplicity of ice.

Flowing rivers brake to a screeching halt in October. In Alaska, this creates flat white highways that last through April, allowing isolated villages to be reached overland. When that happens the landscape metamorphoses into a motionless still life. Thus, for most of the year, the polar realm resembles the desert.

Water's change from liquid to solid requires eighty calories of energy for each gram of ice the size of a sugar cube. Water reaching thirty-two degrees Fahrenheit isn't enough to do the trick, though. Water needs a push, an extra bit of frigid encouragement, to turn solid. Moreover, ice is not a good heat conductor, which means it is a poor cold conductor, too, so it thickens only gradually. To use real numbers, if the air temperature stays at an unwavering four-teen degrees Fahrenheit, studies show that ice will form and grow to four inches thick in two days. That's the minimum recom-mended thickness for ice fishing or other activities pursued on foot.

How much time to double that to eight inches? Not another two days but rather a full extra week. Ice starts fast but then takes a go-slow approach. It requires an entire month more to achieve the fifteen-inch thickness that can support the weight of cars.

Resembling souvenir snow globes, the scene in Fairbanks can look Christmassy even in May and September, for the city has just three snow-free months. But here and everywhere else, an odd cloud dance has to happen for snow to materialize. Water droplets don't simply freeze just because the temperature falls below thirty-two degrees Fahrenheit. First, several water molecules have to col-lide before a potential ice structure can begin to form. *A single molecule cannot freeze.*

Second, if the droplets are pure water, the ice-crystal process is reluctant to get under way at all. It won't happen anywhere near the freezing mark. As if bureaucratic red tape is gumming up the process, ice won't form unless the temperature reaches seventy-two degrees below the freezing point. Forty below zero.[1] So for ice or snow to materialize at a more reasonable and common temperature, the cloud's droplets need a seed or nucleus around which to grow. Since air normally contains lots of tiny floating debris, this is usually no problem. But you'd never guess what the best ice-generating specks might be.

Germs! A droplet readily freezes into a crystal around a living airborne microbe, a bacterium, at any temperature below twenty-eight degrees Fahrenheit. They'll form around a tiny speck of floating clay (kaolinite) a bit more reluctantly and only if it's colder than twenty-five degrees Fahrenheit. And if all they have are specks of silver iodide, the compound used in cloud seeding, they'll start to make crystals below twenty degrees Fahrenheit. But germs are the most common snowflake starters and lie at the heart of 85 percent of all flakes.[2]

So next time you gaze at a lovely snowstorm, inform your favorite germophobe or hypochondriac that living bacteria sit shivering in most of those untold billions of flakes. Then hand him or her a snow cone or organize a catch-a-snowflake-on-your-tongue party.

Once the ice-forming process is started, more molecules join the party, and the crystal grows. It can ultimately become either a snowflake or a rough granule of ice called by the odd name *graupel*. A snowflake contains ten quintillion water molecules. That's ten million trillion. Ten snowflakes—which can fit on your thumb tip—have the same number of molecules as there are grains of sand on the earth. Or stars in the visible universe. How many flakes, how many molecules fashioned the snowy landscape I was observing as I drove east? It numbed the brain.

The white surface stretching off into the distance was of course cold, but even beneath it the ground is permanently frozen here. This *permafrost* is ubiquitous north of the Arctic Circle, just over a hundred miles from Fairbanks. In the southern two-thirds of

Alaska it's common yet spotty. It may not start until a depth of thirty feet in some places, while a few yards away the permafrost lies at the surface.

Residents have no choice but to build their homes, roads, pipelines, and schools on this permafrost. The results are often disastrous. A drive along many Alaskan roads reveals houses sagging pathetically. Inside, floors slant so dramatically that one can almost slide from the bedrooms at either end toward the kitchen in the middle. On this trip, one dispirited native lamented the astronomical costs he was facing, first to jack up the entire building and then to try to create flowing air beneath it so that the frost can reestablish itself and remain year-round. The problem is unpredictability. The ideal technique—building on stilts so that cold air ventilates beneath—usually preserves the permafrost and keeps the house level.

Throughout this vast state in summer, the top few feet of permafrost melts, but the water has nowhere to flow. This creates billions of stagnant pools of various sizes that are perfect mosquito breeding grounds. It's a nightmare. Slap your arm on any random spot between May and August and you will kill ten mosquitoes.

The slow-motion drama of heaving and settling ice and the homes built on them plays out all across Earth's frozen wastelands. Meanwhile the weight of countless snowfalls typically compacts everything below them into ice. In places, this cobalt-blue ice remains rock-hard for tens of thousands of years. We know this from analyzing bubbles of trapped air, which reveal the contents of our atmosphere as it was long before humans made fire or untold domesticated cattle belched methane.

Meteorites plow into the snow, become entrapped, and emerge only when that ice layer has completed its mysterious slow journey down, sideways, and eventually back up to the surface. In the vast parts of Antarctica, where ordinary terrestrial rocks have no business lying on the snowfields, researchers in snowmobiles enjoy gathering up any solitary stones they find, knowing they've probably just snagged a valuable visitor from space using this no-brainer method. The famous black Antarctic meteorite ALH84001, which originally came from Mars, lay conspicuously on the snowy surface of the Allan Hills region when a snowmobiler found it in 1984. It had resurfaced sixteen thousand years after its impact and initial burial, having been subjected to remorseless icy, hydrolic cycles of pressures, releases, and movements in various directions unchronicled and unknowable by anyone.

Slow motion is ice's solemn oath. Even its beginnings are leisurely, since snow typically falls at three miles an hour—the same speed a person walks. But if the snow compresses over a glacial field, it naturally partakes of that ice sheet's even slower creep to the sea. These flowing ice rivers move anywhere from ten feet to one hundred feet a day, depending mostly on the slope of the land. Typical glaciers move a foot an hour, just barely too slow to notice.

Yet beneath much of Alaska's immobile snowy landscape is lively movement. It's an entire biological world. This is the *subnivean* realm.

You don't need to be in Alaska to experience the subnivean universe or even to fall in love with the word. In most of the United States and Europe, the winter landscape—so seemingly motionless—hides constant animated activity on the part of small mammals, including voles, mice, and lemmings. They not only adapt to the snow cover but also rely on it for their very survival. They scurry along wide corridors an inch or two high in the gap between the ground and the bottom of the snowpack, a gap formed after the snowpack contracts a bit.

This region, illuminated diffusely from above in seemingly perpetual twilight, enjoys an air temperature around the freezing point once the snow cover is more than a half foot thick, thus creating insulation against the frigid air above. This subnivean system of open spaces and tunnels lets these mammals move unseen by

Everything seems motionless in this winter scene. Yet under the snow, in a gap between its lower boundary and the ground, lurks the subnivean realm, where small mammals are engaged in constant activity.

many predators, although foxes and owls can hear the scurrying and usually know exactly where to pounce.

Where the spaces between ground and snowpack are filled, the rodents build meandering tunnels, half in the ground and half in the snow. These trenches dominate the scenery as soon as the snow melts down to an inch or two in the spring. Zoological winter motion unfolds continuously, even if our eyes see none of it.

The world's warming climate doesn't help make life in Alaska any easier for the natives, human and otherwise—although you would think the opposite would be true. Temperatures are rising far more at high latitudes than anywhere else. Permafrost, still a dominant aspect of the far-northern experience, is now undergoing dramatic changes. This directly affects the people who live in Arctic villages and communities. According to a recent report from the UN's Intergovernmental Panel on Climate Change, the current steady thawing of Arctic permafrost is "likely to have significant implications for infrastructure, including houses, buildings, roads, railways and pipelines."

Experts believe that by the mid-twenty-first century permafrost will decline by 20 to 35 percent. I could vividly see this happening as I drove toward Chena, passing those sagging homes with funhouse ramps. Even the famous Alaska Highway, called the Alcan by everyone there, is suffering collapse in spots because of the melting permafrost upon which it's largely built. On July 23, 2012, on the seventieth anniversary of the completion of that historic road, the *New York Times* noted, "As the climate warms, stretches of permafrost are...melting—leaving pavement with cracks, turning asphalt into washboard and otherwise threatening the stability of the road."

Where the permafrost has been replaced with seasonal thawing and refreezing cycles, bizarre phenomena materialize. One of them has the odd name of *pingo*. This is a mound up to ten stories tall formed when material on the ground is periodically pushed skyward by the relentless annual heaving.

Some lakes in Siberia abruptly vanish, thanks to *thermokarst* conditions, which occur when rising temperatures melt enough of the permafrost to open up creepy passages into the bowels of the earth. Suddenly large lakes drain away overnight. They empty as if someone pulled a bathtub plug. You can't make this stuff up.[3]

Pulling into Chena after a ten-year absence, I appraised this end-of-the-road, escape-from-civilization outpost anew. It had changed little. It's now slightly spiffed up, thanks to the fact that sellout crowds of Japanese tourists now regularly visit the northern lights, which to them are sacred. But truth to tell, Chena's layout and cabins are still just a notch up from "funky." Indeed, much of rural Alaska is funky. Most villages look like glorified trailer parks. They're dominated by tiny ranch houses or cabins with makeshift steps and small windows and yards piled with old engines and tarps. The sheer unpretentiousness has, I suppose, its own honest appeal.

The last time up here I'd rented a plane, taken advantage of Chena's snowy runway, and headed north past the nearby Arctic Circle to land at villages like Bettles, population twelve, that have no roads at all. When the frozen rivers melt, everything gets in and out by small plane. After I landed on the white-packed runway, pretending I was a bush pilot even though the real McCoys have nerves and guts totally out of my league, I located the single diner-type restaurant among the small hunkered-down cottages. All the patrons looked up from their meals. They kept staring; they don't get many visitors. I was the entertainment. The women were flirty. The men were strangely silent and wide-eyed. Yet late winter is a fascinating time to come—better than summer, with its relentless clouds of mosquitoes and endless, sleep-inhibiting daylight, which means no possible chance to see the legendary lights. "March is the best month," I'd heard the natives say repeatedly.

Chena offers basic wood cabins, dogsledding, and—the big draw—a natural hot spring that feeds a hot pond. You relax in the three-feet-deep, 102-degree water under the auroras while your hair freezes solid. Afterward, a thirty-minute trip in a tank-tread arctic vehicle resembling a giant bulldozer with an enclosed cabin on top takes you up and up, far from the hot springs and the cabins, to a flat region at the top of a mountain surrounded by distant, jagged, snow-covered peaks in all directions. You first put on all the warm clothes you own—two layers of long johns, pants and hoodie, the works, and follow this with an orange government-issue polar jumpsuit. It's still bitter cold.

Last time I was here, when the temperature was minus thirty-five degrees Fahrenheit, I stepped outside with a cup of boiling water and tossed it in the air. The liquid made a loud tinkling, crackling sound and hit the ground as frozen pieces of ice. Tonight felt no warmer, though the thermometer registered a mild minus twenty degrees Fahrenheit.

At the summit, the northern lights not only filled the sky but also painted the snowy landscape green. For a hundred miles in all directions the pointy mountains glowed emerald. This was routine for the hot spring's owners, who bravely took over what was a struggling state-owned enterprise in 2000. Tonight they stared at the waving jade curtains for the fiftieth time this season, once again standing mutely in awe. At least I think it was awe. It might simply have been too cold for small talk. Being in awe and being frozen produce similar behavior.

No one spoke while the aurora undulated. The blotches, rays, arcs, and curtains rustle leisurely, like draperies in some vast celestial kingdom. The changes resemble the mutation speed of low clouds on a summer day. Keep staring, and the movement is barely perceptible. Look away for a minute and then turn back, and the scene has been totally transformed. Here in central Alaska, observers often gaze from directly beneath those curtains so that their "folds" angle vertically up, converging like railroad tracks straight overhead. Blotches vanish and are replaced. Pink fringes come and go. The slow-mo choreography can't be predicted.

But the unseen entity fashioning it, the wizard behind the auroral curtain, is anything but sluggish.

The drama begins with an epic blast of solar particles escaping the clutches of both the sun's gravity and its magnetic field. It was physicist Gene Parker who first surmised, in the 1950s, that the sun, the nearest star to Earth, leaks a continuous stream of broken atom fragments—an outflow he called the *solar wind*. The reward for his prescience: people openly scoffed. It was only when spacecraft launched after 1957 actually detected this nonstop swarm of material—about ten particles for each sugar-cube volume of space, all hurtling outward at a few hundred miles a second—that Parker went from goat to prophet.

Accompanying his promotion came a slow recognition of the ways in which the sun's wind has been affecting our solar system all along. Soon everyone, with the groupie wisdom of hindsight, said,

"Of course! That must be why comets' tails always point away from the sun. Comets are like wind socks. They're blown back by the solar wind. We should have known!"

It took until the 1970s, however, before researchers discovered the truly superdense, superspeedy solar winds that make Parker's solar wind seem by comparison a mere zephyr. These explosions blast away ten billion tons of material at a time at five hundred miles a second. Called CMEs, or coronal mass ejections, they are the real deal and can inflict serious damage on our electrical grid and satellites.

That's the rough motion picture of the sun's particle geysers. But, as always, the devil is in the details. Our planet's magneto-sphere can direct those shotgun pellets of solar detritus safely around and past our world, but only if the swarm's field and our planet's field share the same magnetic polarity—if, for example,

Though their motion is imperceptible to the eye, these Alaskan glaciers typically advance toward the sea at the speed of a foot per hour.

both fields have their norths aimed upward. As they say in magnetic gossip circles, "Like repels like."

Conversely, if the buzzing swarm of solar hornets has a polarity aligned opposite ours, it will transfer its energy to our planet's field. Then the charged particles slither angrily down our magnetic field and into our upper atmosphere. This creates huge electrical charges. Oxygen atoms in the thin air one hundred miles above us have their electrons excited. They emit an alien green glow as these electrons fall back into their preferred, accustomed positions. That's the entire aurora story.

The whole thing is a motion demo. Sun material in motion. Our own air's electrons in motion. The aurora itself, like living abstract art, in vivid, fabulous motion — even if it takes a minute before the scene has completely transformed itself.

What's surprising is how few people in Alaska understand the process even in a general way. They look up routinely at the lights, but I've overheard many "explain" to their companions that it's sunlight reflected off the oceans from the bright side of Earth or deliver some similarly discredited nineteenth-century elucidation.

Apparently, like Gene Parker's supersonic solar wind, science knowledge has its own motion. And this, as in the old days of Jack London and his Inuit fantasy tales, sometimes moves with all the alacrity of Alaska's blue ice.

CHAPTER 7: April's Hidden Mysteries

Deciphering the Secrets of Spring

> *(but*
> *true*
>
> *to the incomparable*
> *couch of death thy*
> *rhythmic*
> *lover*
>
> *thou answerest*
>
> *them only with*
> *spring)*

— E. E. CUMMINGS, "O SWEET SPONTANEOUS" (1920)

It was April, after a warm winter. A million-ring circus erupted in the mountains of the Northeast, the calendar thrown out the window. Bees circled wildly around neon-yellow forsythia, weeks ahead of schedule. At the renowned Cornell University Cooperative Extension, top botany, zoology, and entomology experts scratched their heads, helping local farmers figure out how such early springs affect apple trees and such.

Even after a normal northern winter, with its unchanging monochrome, the noun *spring* doubles as a verb. Countless actions spring to life. They set children's minds going: How fast do flowers open? Trees grow? Insects fly? Saps flow? How does it all happen?

Mere recitations of speed data wouldn't do justice to the whole choreographed enterprise. Not when colorful complexity pops everywhere like jack-in-the-boxes, spurred by sun and warmth. For biology is really the friendly face of physics. As temperature increases, so do all the enzymatic, mitochondrial, glucose-transfer, and other reactions upon which life is based.[1] We mammals make our own heat, and when that's too difficult we hibernate, drop our body thermometers by ten degrees or so, and wait it out. During hibernation a chipmunk's heartbeat slows from 350 beats per minute to as little as four. The pace of activity—within bodies and around them—becomes glacial, as a sleeping community of bears, bats, ground squirrels, and woodchucks snore unseen, often much nearer to our bedrooms than we imagine.

But plants and invertebrates can't do this. They need winter to end. So when it does, and when they—meaning insects, worms, tadpoles, and the like—emerge, so do their predators: birds, raccoons, and foxes. The whole awesome production materializes together. That's why spring isn't just a season. It's a motion-based event.[2]

In subtropical Florida, Southern California, and Texas, spring begins in February and moves one hundred miles north each week. Air travelers observe its vivid border as opening blossoms and leaves rush poleward at the speed of one kilometer, or 0.6 miles, per hour. Spring travels about the same rate as a parent pushing a stroller.

Over the course of three months, spring proceeds more than a thousand miles to envelop all of Maine, North Dakota, Montana, and Washington State as well as parts of southern Canada. In mountainous regions it sweeps first through the valleys and then climbs ever higher up the hills.

Plants bloom in the same sequence year after year. The earliest—snowdrops and crocuses—pop up where the snow has barely melted. They're soon followed by bulbs, such as tulips and daffodils. At that point the changes are measured in days, with the arrival of the bright yellow forsythia bushes. And then the first tree buds and baby leaves burst forth from the impatient chartreuse performers, such as willows, magnolias, maples, and rhododendrons. Cherry blossoms appear around this time, too.

Insects emerge from their winter dormitories. Like plants, they do not wait for a particular date but respond to warmer temperatures. Some, like butterflies, go through the full spectrum of life stages in tree hollows and crevices during the winter—larvae, eggs, and adults—so that they can hit the ground running in the spring. Migrating aviators, such as robins and red-winged blackbirds, arrive earliest to get in on this first action. They catch the initial emerging insects and worms after using the same flyways they traveled in the autumn, when they went south. Now they lay claim to their breeding and feeding territories.

Insect development only occurs when the temperature rises above a particular threshold, which is often fifty degrees. Once it gets warm enough, it almost seems as if spontaneous generation is at play: they're suddenly everywhere. Ants start walking, at an average speed of one-fifth of a mile per hour. (By contrast, thunder covers that same one-fifth of a mile in a second, making thunder 3,600 times faster than ants.)

Each species has its own story and public relations image. Everyone loves butterflies, which boast mellifluous names in all Romance languages: *papillon* in French; *mariposa* in Spanish. Even German manages to make *butterfly* a bit less guttural than usual: *der Schmetterling*. Bees and dragonflies get good press, too.

But not mosquitoes, of course. They come in some 3,500 known species—new ones are still being identified—and have been called the most deadly creatures on earth, thanks mostly to the three varieties that carry the diseases malaria, dengue, and yellow fever. Only the females suck blood from vertebrates such as ourselves.

Movement plays a key role in our mosquito experience. They need standing water and rarely venture more than a mile from their breeding ground. So if you control the stagnant pools of water in your area (meaning no old tires, sagging tarps, and the like) you may be able to completely prevent their appearance. In deeply wooded places, such as Maine and especially Alaska, where small nooks, ponds, and saturated soil left over from rain, melted snow, and permafrost are virtually everywhere, it's a hopeless cause.

Mosquitoes live everywhere on earth except Antarctica and can be so thickly ubiquitous that each member of an Alaskan caribou herd typically loses a pint of blood a day. Despite mosquitoes' prevalence, males only live a week; females a month at best. Their egg, pupa, and larva stages together last just a couple of weeks. So if and when breeding grounds dry up, mosquitoes vanish a month later.

Frustrated at trying to swat them? Scientists who study insect speeds conclude that they seem faster than they really are. This perception issue harks back to that old business of how many body lengths something moves per second. Mosquitoes most often fly at 2.5 miles an hour, so they can't even keep pace with a jogger. But since this translates into 170 mosquito body lengths per second, they may seem supersonic.

Bees routinely move at jogger speed—seven miles per hour. Of the emerging springtime insects, flies, which look the fastest, *are* the fastest, at ten miles per hour. Of these, the horsefly is the champ, as everyone knows who has tried to dodge those creatures from hell. They can fly at 14.8 miles per hour, so only fast sprinters can hope to outrun them. The very quickest insects, however, are the good-guy dragonflies, which appear in 5,680 species and have been clocked at an amazing forty miles per hour. Best of all, they love to eat mosquitoes and have the velocity to catch them effortlessly.

May is when the flower petals of rhododendrons and azaleas add their color, along with crabapple trees and flowering dogwoods,

Cornus florida. Also in May, wisteria blooms, accompanied soon thereafter by the magical lilacs—or at least the most cultivated variety of lilacs, *Syringa vulgaris.* Their heavenly scent, following the magnolias by a couple of weeks, fills the countryside.

Aromas themselves have their own tricky motion, since they can only move with air. Dead calm means that scents scarcely migrate from blossoms. On the other hand, too brisk a wind, and their molecules are diluted and whisked away.

It is still officially spring in early June, when perennials explode almost in unison, along with flowering shrubs such as bridal wreaths and roses and viburnums. The early frenzy is now replaced with the steady rhythms of shrubs, flowers, and trees destined to peak at their own predetermined periods. By the time spring ends, at the June 21 solstice—which is more frequently happening on June 20 as this century progresses, one result of the four-hundred-year Gregorian calendar cycle—the final holdouts, such as hickories and the slowpoke catalpas, have come into leaf even in their northernmost ranges.

This dynamic simultaneous animation of millions of insects, plants, and animals within a hundred yards of your rural home repeats every spring in the same sequence. But now look closer, to the motions hidden behind the curtain.

In 1663, the British philosopher and natural scientist Robert Boyle wrote, "There is in some parts of New England a kind of tree... whose juice that weeps out of its incisions, if it be permitted slowly to exhale away the superfluous moisture, doth congeal into a sweet and saccharin substance." So indeed, one heralded early-spring marker in the northern states is the tapping of maples for the purpose of collecting sap, which is then boiled down for syrup. Since it takes forty gallons of sap to produce a single gallon of maple syrup, copious fluid is required. Many people imagine that all trees have sap running in them during the spring, but it isn't true. Very few emit sap when punctured, and the maples do so only under odd conditions—and only *before* they come into leaf.

Maples produce sap during periods of cold nights and warm days, a situation that occurs most often in March and April. Sap flow stops if the temperature is either continuously above or below freezing and when the nights no longer fall below freezing. It's very odd behavior. You can tap willow, ash, elm, aspen, oak, and many others and you'll never collect a drop of sap. We now know that the reason has to do with freezing inside the tree and then the subsequent warming. This releases expanding gases that push the fluid. And yet no one understands why a sweet, sucrose-filled liquid is necessary or what this has to do with living tree cells. So it's still largely mysterious, though syrup's gooey, ambrosial presence on pancakes temporarily erases any frustrations we might have with science.

By contrast, other trees have saps running upward through their xylems when they are in leaf, and it is not sweet. Plants and trees transpire, meaning water evaporates from their leaves. This creates a partial vacuum that pulls water up from the roots. You'd think the sap would run fastest on hot afternoons, since plants transpire three times faster at eighty-eight degrees than they do at seventy degrees. But sap speed is fastest in midmorning, even if it continues all day.

Superman's X-ray vision would observe that sap is no slowpoke. For years, measurements have been attempted by means of injected dyes and radioactive monitoring, but the past decade's favored method involves thin temperature probes inserted into the tree in various places and the introduction of heat at the trees' bottoms. These show that the rising sap carries the introduced heat upward at rates as fast as one-third of an inch per second. This may not sound fast, but it translates into ninety feet an hour, letting even the tallest trees quickly deliver water from roots to leaves. Most trees are not so speedy, however, with figures closer to eight feet per hour—still sprightly enough for us to see water motion if we could peer through the bark.

In the deep woods, meanwhile, wildflowers push their shoots above ground to take advantage of the preciously brief period of forest-floor sunlight before tree leaves shade them.

As the thermometer climbs, so does the noise level, for sound is the auditory manifestation of movement. A familiar example of this is the chirping of crickets. Only male crickets stridulate, but what's obvious in every rural zip code is that the pace changes with the temperature. The chirping sound, which comes from the top of one wing being scraped along the bottom of the other, gets more frenetic on warmer nights.

Once again, it's the same principle as an old battery failing to start the car on an icy morning. Chemical reactions speed up as temperature increases, as do the metabolic processes in insects, which is why it's always wisest to dislodge an unwanted hornets' nest on a frigid night, when they are too cold to respond. Ants, too, walk at a speed that depends on the temperature. All insects rely on myriad chemical reactions in their bodies and have no way to speed them up except to hope for an environmental warm spell. As the temperature rises, they more easily reach the energy threshold necessary for chemical reactions that will let them perform various muscle contractions, a prerequisite for walking, flying, or—in the case of crickets—chirping.

The rate at which crickets chirp depends also on the species, but a good average is about a chirp a second when the night air is fifty-five degrees. If you want to be a show-off at your next scout meeting or Trivial Pursuit game, you can tell everyone that the name for the relationship between temperature and chirping is *Dolbear's law.*

Amos Dolbear, born in 1837, was *almost* the world's most famous person. And not because of insects. When we think of the invention of the telephone, radio, and electric light, the names Bell, Marconi, and Edison spring to mind. But for a whisker of chance it would have been—some say it should have been—Dolbear alone.

He was no toolshed tinkerer. Amos Dolbear graduated from Ohio Wesleyan University and ultimately became chairman of the physics department at Tufts University. While still in his twenties he created a working telephone that he called a talking telegraph, a device that used his own receiver constructed of a permanent magnet and a metallic diaphragm. This was in 1865, fully eleven years before Alexander Graham Bell patented his version of the telephone. Later, Dolbear tried strenuously to show that he and not Bell was first, and the case went all the way to the United States Supreme Court. The journal *Scientific American* reported on June 18, 1881: "Had [Dolbear] been observant of patent office formalities, it is possible that the speaking telephone, now so widely credited to Mr. Bell would be garnered among his own laurels."

Defeated but still energetic, Dolbear turned to wireless communications, and in 1882, while a professor at Tufts, he succeeded in sending signals a quarter mile using radio-wave transmission through the earth. Made wise by his bouts with Bell, he filed for and received a patent for his "wireless telegraph," improving its transmission capability to a half mile by 1886. This was groundbreaking and beat out the theoretical work of German physicist Heinrich Hertz and, by a full decade, the practical inventions of the Italian Guglielmo Marconi. Dolbear's patent later prevented Marconi's company from doing business in the United States and forced the Italian to purchase Dolbear's patent.

Dolbear even invented a system of incandescent lighting ahead of Thomas Edison, but here, reverting to earlier form, he didn't pursue it fast enough to edge out Edison's later monopoly. In short, he was an eyeblink from going down in history as the inventor of all the most important technologies of our time.

It seems that none of these inventors actually stole from the other. Rather, in a strange echo of nature's predilection for patterns, the same ideas occured to different people at around the same time—a sort of hundredth monkey effect that seems to happen more often than random chance would suggest.[3]

Out of left field, and bearing no relation to applied physics, Amos Dolbear suddenly submitted an article that was accepted for publication in the November 1897 edition of *The American Naturalist*. Titled "The Cricket as a Thermometer," Dolbear's article spelled out the connection between the night's temperature and the rate at which crickets chirp. The formula he expressed became known as Dolbear's law, which still remains widely known in esoteric entomology circles. You simply count the number of chirps that occur in fourteen seconds and add forty. Voilà: You get the current temperature in degrees Fahrenheit. This assumes you're hearing the snowy tree cricket, the most common variety in the United States.

Fame has fully eluded Amos Dolbear a century after he left this planet. Perhaps we can remedy this, just a little, by announcing the temperature during our next camping trip while grandly invoking Dolbear's law.

Crickets easily catch our notice because we humans are very aware of repetitions that are roughly in sync with our own heartbeats—and crickets' stridulation rate rarely diverges by more than 50 percent from this. We especially notice things that repeat between 0.5 and ten times a second. Slower than that and we may regard the individual events—such as the hooting of some owls—as independent and not link them into a single activity. Faster than that and we perceive them as a steady sound, its own sole event rather than an assembly of others.

For example, many mosquitoes give off an annoying drone in the musical note A, the same as a telephone's dial tone.[4]

It's caused by wings flapping at 440 beats per second. But other mosquitoes flap six hundred times a second, producing something like a D or D-sharp. In either case our ears perceive no sensation of separate mosquito beats. Anything more than about fifteen beats a second seems a single tone.

Meanwhile, as bees jerkily dart through the air to pollinate trees and flowers, their low, buzzing pitch comes from wings flapping 230 times a second, the note of A-sharp one full octave below the mosquito drone. But frogs and salamanders are ready for a variety of flying insects, as they quickly arise from their hibernation and start to fill the air with mating songs.

Above all the marshy melodramas, fireflies blink on and off. Their bioluminescence, caused by the enzyme luciferin interacting with oxygen, typically emits a yellow-green light in the same color as an aurora.[5] And, like the northern lights, fireflies produce radiance without heat. Also like an aurora, fireflies produce light that is unreliable. The insects are only active for a few weeks in late spring and summer, only when the night is warmer than fifty degrees, and—for reasons that remain mysterious—they almost never turn on their lights west of Kansas.

As spring progresses, the season's new crop of young mammals becomes obvious to country dwellers. We see bear cubs and fawns staying close to their mothers, but rarely do we observe the more secluded, furtive animals, such as coyotes, who have their pups then, too. None of these large mammals actually breeds in the spring. They mate during the previous fall, instinctively planning for their young to be born during spring's food festival. Mostly it's the small mammals who are going out on dates during the spring, and even they time their activities to catch the peak of the season's abundance. Chipmunks start to be active enough to breed as early as February and are thus among the first mammals we see, even when patches of snow still prevail. They rely on having multiple entrances to their dens to evade predators, and they protect themselves with their jerky speed.

It's often not enough. Despite some silly claims on the Web that various rodents can whiz along at thirty-five miles per hour, actual laboratory track tests and field measurements show that rodents have a top speed of about ten miles per hour, give or take a couple. They may seem much faster because, once again, they traverse many rodent lengths per second. But a mouse can only dash at eight miles per hour. The common eastern gray squirrel can hit twelve miles per hour on a good day. Unfortunately for them, they cannot outrace their usual predators if the match is held at a straightaway. A house cat can run more than three times faster than any mouse. It's not a fair contest between Tom and Jerry.

WHO CAN CATCH WHOM

The Chase Is On:
Top Speeds of Common Mammals

In Miles per Hour

Chipmunk	7
Mouse	8
Squirrel	12
White-tailed deer	30
Cat	30
Grizzly bear, black bear	30
Rabbit	30
Fox	42
Coyote	43
Fastest dogs	44

The very fastest animals? None races through American forests: it's a tie between the cheetah and the sailfish. Both can reach sixty-eight miles per hour. The fastest-ever racehorse, at least in the 1.25-mile category, was Secretariat. That day in 1973 when he left all other horses in the distant dust while winning the Kentucky Derby, he posted an average speed of thirty-eight miles per hour.

As for winged animals, their speed depends on their motives. Most cruise at twenty to thirty miles per hour, and this is true for small birds as well as large ones. Geese and hummingbirds fly at the same speed. Nearly all birds can tuck in their wings and dive much faster than they can fly should the need arise. Peregrine falcons are

Nearly all birds fly between twenty and thirty miles per hour. But their wing-beat rates vary enormously: it's 1,250 flaps per minute in hummingbirds but closer to one hundred flaps per minute in these greylag geese. *(Michael Maggs, Wikimedia Commons)*

renowned as the very swiftest birds, able to achieve two hundred miles per hour in a dive, although half that is their usual behavior. However, even two hundred miles per hour is perhaps no "achievement": a human skydiver reaches that same speed in a headfirst posture with his arms tucked to his sides. It's the simple matter of terminal velocity; no skill required. A falcon cannot outrace a diving daredevil.

Birds can catch field mice, squirrels, and chipmunks without batting an eye. Squirrels have the most visibly dramatic strategy of defense, routinely zigzagging so that a swooping hawk has a hard time aiming at the fleeing rodent. But birds can and do choose different speeds for different purposes. A hawk on reconnaissance patrol, loitering in the sky in search of prey, would want to maximize her endurance and move her wings leisurely to preserve energy and stay aloft for hours. But a seabird trying to reach a distant hunting ground would want to maximize her range. This doesn't usually entail going fast or even flying long distances through the air; exploiting wind currents might be the key. And birds are sometimes forced to maximize speed, as they do when pursued by a predator.

Suffice it to say that nearly all birds fly between ten and forty miles per hour, and most cruise in the twenty-to-thirty range. Plenty fast enough to catch flying insects, few of which can attain twenty miles per hour.

But more is happening than meets the eye. What we see is, in many ways, not as fascinating as what we could potentially detect with X-ray vision (with which we could peer through bark) or time-lapse perception, because the most dramatic magic unfolding in spring is the act of *growth*.

Trees are classified according to a slow, medium, or fast growing rate. Slow means less than a foot a year. Fast means more than two feet. Medium is in between. Each species is distinct. Sugar maples barely alter their appearance year over year, while willows change shape quickly.

HOW FAST DO TREES GROW?

Fast (≥2 feet / yr)	Medium	Slow (≤1 foot / yr)
Elm	Linden	Balsam fir
Honey locust	Norway maple	Black walnut
Red maple	Scotch pine	White oak
Ash	Red pine	Butternut
Birch	Spruce	Sugar maple
Black locust	White pine	
Box elder		
Cottonwood		
Red oak		
Silver maple		
Willow		

Spring brings the year's fastest growth in trees and plants alike; shoots push up as much as an inch a day. None of it crosses the threshold of visible motion. The plants that come closest are some of the climbing ivies, which use almost spooky holdfasts, and the wraparound climbing stems of wisteria, which can extend ten feet per season and which seem science fiction–like when viewed through time-lapse photography.

Likewise, if we could peer through the ground, we would see snaking roots advancing by two inches to as much as two feet per week. The all-time growth winner, however, is not a plant most of us get to enjoy. It's bamboo. This can emerge from the earth at its full thickness and then head upward at speeds just barely too slow to visually discern. The all-time record is a measured thirty-nine inches in a single day. That's one and a half inches an hour.

So this single season, spring, reliably delivers nature's outstanding hurry-up exhibits. Quick change is what's dramatic, especially when clothed in vivid colors—and *change* is another word for motion.

PART II

THE PACE QUICKENS

CHAPTER 8: *The Gang That Deciphered the Wind*

A Desert Dweller's Airy Spells Last for a Millennium, While Two Oddballs Dodge the Inquisition

Gray-eyed Athena sent them a favorable breeze,
A fresh west wind, singing over the wine-dark sea.
— HOMER, THE *ODYSSEY* (CA. EIGHTH CENTURY BCE)

In the Bible, John 3:8 says, "The wind bloweth...and thou hearest the sound thereof, but canst not tell whence it cometh, and whither it goeth."

Blowing wind had started me on this quest to understand natural motion. Yet being harmed by the wind was hardly a unique experience. It's a familiar scenario in global literature. The specter of an invisible entity that destroys houses has aroused fear through the ages.

But I knew where I must goeth. To the consistently windiest place in the hemisphere. Where anemometers measured the all-time fastest-ever wind gust in a record that stood for more than a half century. That blast duplicated the inside of an EF4 tornado.[1]

New Hampshire's Mount Washington stands for more than mere Guinness-type record holding. Its famous gusts make people itch to experience the wind for themselves. To accommodate them, the state built a road to the summit back when Abraham Lincoln was in the White House. Families looking for a bit of adventure authenticated by a boastful bumper sticker have made the pilgrimage ever since.

Sure, I could lazily get in my old four-seater plane and fly myself over that 6,288-foot peak, but how would that bestow the experience of its famous winds? Besides, I was scared. Wind acts in violent ways around mountains, and Mount Washington possesses an odd configuration that funnels the air with the best of them. I remember reading about a Boeing 707 jetliner flying near Mount Fuji in Japan on March 5, 1966, that, tragically, had its tail torn off by orologically induced turbulence.[2]

Who in ancient times could have begun to understand swirly air? Who could visualize any mechanism by which Earth's five thousand trillion tons of gas are set into perennial motion? None of the ancients tackled the whats, hows, or whys of the gaseous realm. The Westerner who went the furthest was Aristotle, who declared air to be an "element" that liked to rise.

Instead, people asked how moving air could benefit them: What's in it for me? One of the first technological lightbulbs to go *boing* in the ancients' minds was the idea of employing air as a free power source.

Air energy has been harnessed since earliest recorded history, even by a sparse human population that didn't reach two hundred million until the time of Christ. As far back as 5000 BCE, wind propelled boats along the Nile. By biblical times, sailboats were a common sight.

It took an amazingly long time — fully five thousand years after the first canvas sails — before moving air was utilized mechanically. The Chinese did it first, around 200 BCE, when they erected windmills and fitted them with gears that pumped water for irrigation. Soon after, the idea spread to the Middle East, where inhabitants built windmills that had woven reed sails and were geared with a vertical revolving shaft for grinding grain.

The Persians were next in line to use wind power and introduced it to the European regions still under the rule of the Roman Empire by 250 CE. Another few centuries of achingly slow technological progress brought an upgrade to Windmills 2.1, which featured better materials, such as metal gears, and larger, more efficient vanes. These windmills appeared in Afghanistan in the seventh century and Holland by the 1200s. These bigger structures grandly drained marshes and fertilized fields and ultimately even pumped water for American settlers heading west in the 1800s.

For all that, no one seemed obsessed with figuring out what, exactly, is air. Or how far it extended upward, or why it should ever budge. No one guessed that it's a blend of different gases, each with distinct properties. No one puzzled over the bizarre fact that — unlike the sun and moon and the habitual tides and the seasonal rains and the predictable cycles of crops and insects and such — the wind acted capriciously. Sometimes it didn't waft at all, then it could howl destructively an hour later. Strong winds often accompanied thunderstorms. Yet equally ferocious winds could blow from cloudless skies. No other aspect of the everyday environment displayed such wild whimsy.

Even in the early twentieth century, no one knew about well-defined air masses. It wasn't until after World War I, appropriately enough, that the word *front* was coined to describe this novel idea of warring globs of air that produce inclement weather along their boundaries.

The really juicy discoveries started in the eighteenth century and then accelerated in the nineteenth. But a few brilliant thinkers made laudable contributions much earlier.

It was Aristotle in 350 BCE who coined the word *meteorology*, which is simply Greek for the science of "high in the sky." But the study of the atmosphere and its rich, vast, and varied antics perhaps began in earnest a half millennium earlier in India, when the ancient holy texts of the Upanishads were composed. These writings discuss at length the ways in which clouds form and rain is produced and even attribute the phenomenon to the seasonal cycles that result from the movement of Earth around the sun.[3] Around the year 500 CE, Varāhamihira wrote the classical Sanskrit work *Brihat Samhita,* which expounds on complex atmospheric processes such as hydrolic cycles, cloud formation, and temperature transformations attributable to solar heating.

Another half millennium passed with the Western world sound asleep. It was the Dark Ages, when advancements—so promising during the glory days of the Greeks and in ancient India and China—simply stopped cold until the 1500s. Or so we are taught. What everyone forgets are the four wonderful centuries when knowledge was prized in Persia and the Middle East. This was the golden age of Arabic science. While one side was dark, another basked in the sun.

I have a hero from this era. Born in Basra in what is now Iraq in 965 CE, he was Abu Ali al-Hasan ibn al-Hasan ibn al-Haytham, familiarly called ibn-al Haytham in the Arab world. Let's be gentle on ourselves and refer to him by his Latinized name—Alhazen.

He had extensive knowledge of the Greeks and wrote approvingly of Aristotle and disapprovingly of Ptolemy. In what was a groundbreaking approach, he did not merely theorize or speculate but performed careful experiments.

In 1021, Alhazen became the very first to accurately describe the way air bends, or refracts, light. He proved through rigorous observations how the atmosphere creates twilight and said that its very first traces begin when the sun is nineteen degrees below the horizon. Today's modern figure is eighteen degrees.

More impressive—and this is why I applaud him—he was among the first (and likely *the* first person) to use the scientific method to get at the truth. Alhazen employed complex, accurate geometric calculations to determine that the height of Earth's atmosphere is—drumroll, please—52,000 *passuum.*

You're not impressed? That's because you probably haven't lately used that Latin unit of length. It was equal to five feet. Do the math and you'll find that Alhazen's figure for our atmosphere's height was forty-nine miles.

Back then nobody—absolutely nobody—had the slightest clue how far up the air extends. Or whether it ever stops. For all anyone knew it could continue for four miles or four million. Alhazen said forty-nine miles. These days most authorities place the figure at fifty-two miles, the top of the mesosphere. And yet who in the West has even heard of Alhazen?[4]

If Alhazen has any fame at all in the West, it's because he invented the pinhole camera, which I sincerely wish everyone would experience at some point because it's amazing and fun. Once in a while you'll witness something similar when a tiny bit of light enters a dark room through a hole in drawn shades. Exquisite, animated, filmlike details of the world get projected onto the walls and ceiling. It's riveting. Alhazen's fellow desert dwellers must have flipped. Alhazen also discovered the laws of refraction and was able to separate light into its constituent colors. He studied eclipses and optics and correctly figured out the math behind them.

How he had the time for so much study and experimentation is a story he probably loved to share with his analyst. It's a strange tale that began when he still lived in Basra and would read about the Nile's famous annual floods. In a moment of overconfidence he wrote that the river's destructive autumn inundations could easily be controlled by a system of reservoirs and dikes, which might also serve to store the water for use during the long dry season. It was easy for him to imagine such technology, but these innocent published musings unwittingly set the stage for personal-life changes that would have ranked high on the modern Holmes and Rahe stress scale.

When Alhazen arrived in Cairo, the caliph, by all accounts a testy, unpleasant fellow who had heard about Alhazen's claims, summoned him and said, "Okay, do it." Alhazen was taken on a tour of the various floodplains. I wish I could have seen his reaction. He must have blanched. Observing the flood regions in person, the pragmatic Alhazen immediately knew that his plan could not possibly work, not in a million years.

But rather than admit his mistake and take a chance that the murderous caliph would have him executed on the spot, Alhazen tried a risky ploy. Using a technique later perfected by draftees evading the Vietnam War, he feigned madness. He figured the caliph would just have him tossed out on the street and that would be that.

He was wrong. The ruler instead ordered him locked up under permanent house arrest. He was not permitted to experience freedom or mingle with the public ever again.

This good-news-and-bad-news story resulted in Alhazen having ten full years, starting in 1011, to do nothing but immerse himself in the writing of innumerable brilliant treatises, including a notable work on optics that ranks with that of Newton seven centuries in the future. He was finally set free when the caliph died in 1021, at which point he could finally shake off the crazy act that he had probably gotten pretty good at.

The next unfolding of air's secrets arrived more than half a millennium later and involved various aspects that probably should be considered separately. Consider, for example, the pressure, or weight, of the air. Each square inch of Earth's surface is, famously, pressed against by a column of air weighing nearly fifteen pounds. In modern times we experience this in a fast-moving elevator or when an airplane descends. Our ears pop.

We're used to this. But Aristotle, on one of his bad hair days, insisted that air exerts no weight on us at all. And Galileo, normally the iconoclast, meekly accepted Aristotle's incorrect call without argument.

This was the airy mind-set into which Evangelista Torricelli was born, in Faenza, part of the Papal States, in 1608. He's another unsung hero, virtually unknown today, even though he was the one who figured out why the wind blows.

Torricelli lost his father at the age of four. He was raised and educated by an uncle and studied mathematics at a Jesuit college. When he was twenty-four he read Galileo's *Dialogue Concerning the Two Chief World Systems* and wrote to the great man, telling him that he, too, believed in the Copernican sun-centered model. It was a quick way to get on the irritable man's good side.

Torricelli didn't know it yet, but that was a perilous opinion to commit to in writing, given that Galileo was condemned by the Vatican the next year, 1633, and nearly burned at the stake for that very belief. No Jesuit could safely afford to ally himself with such heresy, and Torricelli remained silent thereafter.

The bearded Galileo, soon a prisoner under house arrest, invited Torricelli to visit him, and Torricelli accepted, though it took him five judicious years before he showed up at the door. It was around this time that he started to make scientific breakthroughs in understanding the air and brainstormed inconclusively with Galileo about a very puzzling issue brought up by yet another Italian mathematician and astronomer, Gasparo Berti.

Between 1639 and 1641, Berti experimented with long vertical glass tubes more than three stories tall, filled with water, which had both ends plugged with corks. The bottom end was placed in a pool of water and its stopper removed. What happened next produced the head-scratching.

Some water drained out into the pool, but most of it stayed in the tube. At a height of thirty-five feet — three and a half stories — the water's surface settled a little to leave an empty space in the top of the cylinder. The issue was why the water always remained piled up that high.

The long column of enclosed water always stopped draining when it was thirty-five feet tall, never significantly more or less. Galileo believed the vacuum at the top had enough sucking power to hold up the weight of all that water, the way a milk-filled straw remains full as long as you keep your finger over one end.

But in 1644 Torricelli hit on a different explanation. What if it wasn't the vacuum sucking and holding the water up but rather our atmosphere's weight pressing down on the pond that supported the water? In other words, perhaps the apparatus was like a balance. Maybe it weighed the air above the pond, whose downward pressure was exactly sufficient to keep aloft a thirty-five-foot-high water column. He noticed another oddity, too: the level changed from day to day, going up or down by about a foot.

Berti and Torricelli kept ordering these custom-made, frustratingly fragile four-story glass cylinders and had carpenters build special openings in their homes to let them poke skyward, while others followed their experiments with interest. What the heck did it really mean? A century earlier those suspended water columns would have represented a simple oddity of nature, one of thousands that earned no more than shrugs. But in seventeenth-century northern Italy, nature's foibles had become mesmerizing. They beckoned like some fabled El Dorado, pregnant with the promise of revealing profound underlying secrets.

Those tubes. Those freaking unwieldy glass tubes. They were causing Torricelli's neighbors to whisper "witchcraft." He'd already dodged a bullet with his ally Galileo, but he could still get in trouble. Anxious to end all the staring at his strange through-the-roof straws, and already experimentally filling his tubes with heavier liquids, including honey, Torricelli had a brainstorm for a truly portable device that could be hidden from prying eyes. Because liquid mercury—quicksilver, as it was then called—weighs fourteen times more than water, a tube of that liquid metal could be relatively short and still be useful for his experiments. Torricelli then filled a tube with a mercury column that stood only about thirty inches high and placed it in a pan that was also filled with liquid mercury, thus creating the first barometer.

Its level varied from day to day by as much as an inch, and Torricelli correctly surmised that the weight of air pushing on the mercury pool in the pan must change by about three percent. The way it varied was intriguing, too. The mercury tended to stand highest on cool, clear days and lowest when the weather was stormy.

The French mathematician and physicist Blaise Pascal heard about Torricelli's apparatus in 1646, along with the furor over what, exactly, was keeping those columns of water and mercury standing so strangely erect. Was it the vacuum in the tube pulling or the atmosphere outside the tube pushing down the liquid in the pan or pond? Pascal had a brainstorm, a way to find out once and for all.

If air had weight, there'd be less of it as a person ascended a mountain. Logically, a barometer in an elevated location would display a lower quicksilver column. Was that true? Pascal asked his brother-in-law, who lived in the hills, to perform this decisive experiment. In September of 1648, the height of a mercury column was noted at the base of the Puy de Dôme and then periodically during the ascent to the summit. Bingo: the barometer got lower the higher one went. And not by a little, either. It wasn't subtle. The mercury plummeted a full inch for each thousand feet of ascent. At the top of the 4,800-foot mountain, the mercury stood 24.5 inches tall instead of the twenty-nine inches seen at the base.

Case closed. Not only did Pascal prove the weight of the atmosphere, he also effectively created an altimeter, a way to find one's elevation. Today, newer models using dry diaphragms instead of mercury grace every airplane cockpit.

In 1644 Torricelli wrote the famous line: "We live submerged at the bottom of an ocean of elementary air, which is known by incontestable experiments to have weight." He also soon delivered the world's first scientific description of the cause of air motion: "Winds are produced by differences of air temperature, and hence density, between two regions of the earth."

Torricelli went on to design and build microscopes and telescopes but did not live long enough to gain world renown. Just three years after Pascal had proved him right, he contracted typhoid fever in Florence and died at the age of thirty-nine.

But his invention became all the rage. As we've seen, humans are quick to perceive patterns, and the daily rises and falls of barometers were intriguingly linked with the approach of sunny and rainy weather respectively. It was a forecasting device!

Everyone wanted one. By 1670, many clock makers started producing them for wealthy patrons. A century later, most upper-class homes prominently displayed ornate wooden barometers decorated with magnificent inlaid designs. There were more than 3,500 registered barometer makers in the Western world between 1670 and 1900.

Around 1860, the British admiral Robert FitzRoy, former captain of the *Beagle*, on which Darwin made his famous voyage, started publishing forecasting instructions linked to changes in barometric pressure. He explained newly discovered intricacies. For example, he found that unusually strong barometric highs and lows as well as rapid pressure changes are often followed by wild winds as air frantically tries to go from a high-pressure to a low-pressure region. From that point on, all mariners obsessively consulted barometers before embarking on voyages of any length. Such was the importance of changing air pressure.[5]

Today we're aware of a fascinating cornucopia of pressure events. As you travel to a new location higher above sea level, the temperature falls by around five degrees per thousand feet of climb.

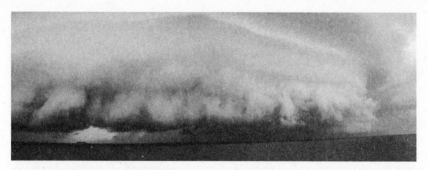

Dwarfing a cruise ship is a shelf cloud of a thunderstorm. Beneath it, winds often hit fifty miles per hour.

This is huge. It means that Denver, up at five thousand feet, is fully twenty-five degrees cooler, on average, than a sea-level city at a comparable latitude.

Thanks to the compression of lower air layers by the weight of everything above, exactly half the atmosphere lies below eighteen thousand feet. A barometer therefore falls to half its sea-level reading when at that height.

Want to feel what it's like up there? You can get close. The highest you can *easily* ascend and still have your feet on the earth is not in Europe or the United States but in South America. I went there in 1988. First you fly into the Venezuelan city of Mérida, snuggled in the Andes in the westernmost part of that country, where you're already a mile high. Then you go to an amazing cable car that takes your breath away as you dangle a zillion miles above nothingness. It vertically climbs an astonishing ten thousand feet—equal to nine Empire State Buildings stacked one atop the other. You go up and up, swaying with every breeze, until you reach 15,629 feet. Now you're on the top of Pico Espejo, a stone's throw from the summit of the famous Pico Bolívar, the highest point in Venezuela, a mere seven hundred feet higher.

In aviation classes they teach pilot trainees that some people feel the effects of altitude at a mere five thousand feet—after all, few small planes are pressurized. Because the blood of people who live or spend a week or more at high altitudes has far more red blood cells than that of people at sea level—unless you are conditioned as well as the Hulk and have a major upgrade over his presumed blood composition—your arrival at Pico Espejo will make you instantly dizzy and maybe euphoric. Walking more than a couple of steps will be exhausting. Here you can perform high-altitude experiments if you can remember what you were doing from one minute to the next. Yet this lofty, picturesque perch in the Andes is still some two thousand feet below the half-atmosphere threshold.

A handful of Himalayan mountaineers have gone way beyond that and experienced even Mount Everest's 29,035 feet without supplemental oxygen. But they, of course, are pretty much space aliens.

Forget about climbing. If we stay with the lazy concept of simply driving as high as possible, there are a few places outside of Leh—in northwestern India, north of the Himalayas—where the pothole-ridden dirt road goes through passes that are seventeen thousand feet high. Still not quite at that magical eighteen-thousand-feet, halfway-to-outer-space milestone. For that there is supposedly a motorable pass called Suge La, west of Lhasa, Tibet, situated at 17,815 feet, and at Semo La, at 18,258 feet, between Raka and Tso-chen in central Tibet. Let me know if you ever do it. You probably will not need E-ZPass.

As you ascend to any new height, wind speed generally increases, and the boiling point of water drops by roughly 1.5 degrees for each thousand feet. It adds up. Sherpas simply shrug and serve tepid tea, because water boils off before it can get very hot.

With the advent of truly high-altitude balloons and, especially, rocket-mounted instruments, the discoveries got weirder. In the 1950s, scientists learned of a frightening point called *Armstrong's line*. (No relation to Neil, the first man on the moon. It is named after Harry George Armstrong, who commanded the United States Air Force's School of Aviation Medicine at Randolph Field, near

San Antonio, Texas, between 1946 and 1949.) Pegged at between 62,000 and 63,500 feet, or twelve miles high, it is *the elevation at which water boils at body temperature.* There, exposed body fluids — such as those in your eyes, your saliva, and any blood outside pressurized veins and arteries — simply boil away. This isn't good for you.

As for high-altitude air motion, during every preflight check when I fly my own plane, I use a great aviation resource, the National Weather Service's Aviation Digital Data Service (http://aviationweather.gov/adds/winds/), to check how the current wind increases at various heights. Right now in Ohio it's calm at the surface. But the wind blows at twenty miles per hour at three thousand feet, whistles at thirty-five miles per hour at six thousand feet, screams at 115 miles per hour at twenty-four thousand feet, and blows a tornado-like 180 miles per hour at thirty-six thousand feet.

That's the jet stream. It's a strange, narrow cylinder of superfast westerly winds that exists on several other planets, too. Its discovery unfolded when scientists watching the skies after the famous 1883 eruption of the Krakatoa volcano saw high-altitude ash whizzing toward the east at tremendous speed. They called this phenomenon the equatorial smoke stream. Then in the 1920s, Japanese meteorologist Wasaburo Oishi detected identical high winds going eastward from Mount Fuji and released balloons to track it. But it took World War II aviators to confirm that, indeed, if your plane runs into a jet stream, it can speed up by as much as two hundred miles per hour. This helps explain why a coast-to-coast flight in the United States is an hour shorter when heading eastward. The jet makes the trip using 20 percent less fuel. Strangely enough, airlines don't offer a discount for flights in that direction.

I finally reached the Presidential Range, in northern New Hampshire, one of the least populated areas east of the Mississippi. I turned into the entrance to the park and paid my fee. There I learned that one's car had better be in good shape to make the climb up Mount Washington, and the brakes better be good enough to make the unrelenting descent, and some models were simply not allowed: you had to have an operational first gear, for example, which eliminated some Lincoln Continentals. But I did, so I stepped on the gas, and my Solara convertible whined against the incline, tackling a road built in 1861.

I did not hit the mountain at the right time. I had no chance to witness people being blown off their feet. In August, when I was there, the summit's average wind speed of twenty-four miles per hour is only about half what it is in January, which is when things get crazy. Five of Mount Washington's Januarys have seen gusts higher than 170 miles per hour—the same as the strongest-level hurricane. But it's never happened in summer. This was a place of extremes, but where was my dramatic story?

A hurricane-force gust blows a scientist off his feet at the observing station atop New Hampshire's Mount Washington, the windiest place in the Northern Hemisphere. *(Mount Washington Observatory)*

I arranged interviews with Mount Washington's scientists, who live for eight days at a stretch in the observatory atop the summit. I was seeking specific information, fishing for an exact blown-off-the-mountain wind speed, the kind of juicy stat that conveys a dramatic image. But with the cautiousness of a good meteorologist, Dr. Brian Clark wouldn't give me one.

"There is no threshold wind velocity that will reliably knock people down. It depends on a person's height and build," he explained.

"Well, what wind speed will blow *you* over?" I asked.

"It depends. It's much harder to stay standing when it's very gusty as opposed to a steady wind that you can lean into."

"How gusty?"

"It depends."

I wasn't getting anywhere. I tried a different ploy.

"Listen, your own media relations person, Cara Rudio, already told me that most people are knocked off their feet when gusts hit the high eighties or low nineties. Would you agree with her?"

"She said that?"

"Yes."

This gave Clark some pause. He then insisted that experienced professionals, who venture out each hour to clean ice off instruments and take readings, often remain on their feet even above one hundred miles per hour. After all, he explained, the entire staff undergoes "slide and glide" training.

"Well," he finally and grudgingly conceded, "I guess no one could remain upright at one hundred and fifty miles per hour."

Why is this place so windy? It seems Mount Washington sits at a *perfect storm* location at the convergence of three major storm tracks, plus its prominence within the surrounding landscape amplifies the winds, plus there's a funnel effect, like the venturi of a carburetor. In terms of the world's highest-ever gust observed by people (as opposed to unmanned instruments), Mount Washington still holds the record: 231 miles per hour, recorded in April of 1934.

* * *

After Evangelista Torricelli proved that air moves in response to differences in pressure and temperature, there still remained the small matter of what, exactly, air is. This required more than a full additional century of labor. The knowledge arrived via a flurry of discoveries just before the American Revolution.

Turns out air is a simple mixture of around 78 percent nitrogen and 21 percent oxygen. Everything else is an afterthought—less than 1 percent combined. And of that remaining 1 percent, argon—the inert gas inside every lightbulb—constitutes 0.93 percent. Nitrogen, oxygen, and, okay, let's include argon. The big three. Now you've identified 99.93 percent of the atmosphere. (When it's dry, that is. The presence of water vapor varies so much from place to place that it's usually omitted in this kind of conversation.)

After argon, you're down to tiny fractions of eolian ingredients, such as carbon dioxide, a mere twenty-fifth of 1 percent. It's scarcely present at all, despite its greenhouse notoriety. Yet it was discovered before the other gases.

That's because CO_2 is readily emitted in all sorts of chemical reactions, such as the one that happens if you throw some baking soda into vinegar. It's easy to produce, hence it was easy to discover. Air's two major components were a bit trickier yet were isolated almost simultaneously. Nitrogen was identified in 1772, oxygen in 1774. Their main distinction was immediately obvious. One supported life and combustion, the other didn't.

The big nonoxygen player soon acquired a ghoulish reputation. Nitrogen's discoverer, Daniel Rutherford, called it *noxious air.* Other chemists alluded to it as *burnt air.* The French "father of modern chemistry," Antoine Lavoisier, called it *azote,* from the Greek *azotos,* meaning "lifeless." Mice placed in it quickly died. But officially designating the bulk of Earth's atmosphere lifeless would have been kind of creepy. Fully eighteen years after its discovery, the current name was suggested.

As for oxygen, this was the precious, life-sustaining element everyone was then trying to isolate. Because—unlike the "introverted" nitrogen—it eagerly combines with most other elements, it makes up two-thirds of our bodies by weight. By itself it accounts for half the mass of the moon. When coyotes react to the lunar crescent, it is basically a display of oxygen howling at oxygen.

CHAPTER 9: *Blown Away*

A Fanatical Mariner Takes the World
to the Edge of Violence

Will the wind ever remember
The names it has blown in the past...
—JIMI HENDRIX, "THE WIND CRIES MARY" (1967)

In the same year the mostly-made-of-oxygen Joseph Priestley discovered oxygen, Francis Beaufort was born in Ireland. Thus we now finally arrive at the modern study of airy motion to which his name was attached for centuries, thanks to the famous Beaufort scale.

For it is one thing to finally know why the wind blows and what it is. It is quite another to watch houses carried off, as in *The Wizard of Oz*. Why on earth should moving air ever accelerate from a sixty-mile-an-hour gale that merely rips branches off trees to a two-hundred-mile-an-hour fury that kills dozens at a time?

Classical and Renaissance scientists had obsessed over air and triumphed. But it took until the nineteenth century for a new breed of scientist to emerge, one whose fascinations revolved around *violence*.

When the year 1971 began, the world had no system for measuring or even speaking about the most savage winds. The Beaufort scale, which we will explore in a moment, had been used for 166 years

130

but only goes up to "hurricane." Once your roof blew off you were on your own to shout out any further wind-defining expletives you wished to add.

In reality, hurricanes are no more alike than are earthquakes: that single word can mean everything from an unfelt tremor to a violence that literally tosses animals in the air and abruptly kills a half million people. Some hurricanes let intrepid reporters deliver televised accounts safely from the boardwalk; others can blow that same meteorologist to the ground.

So Bob Simpson, director of the National Hurricane Center, teamed up with civil engineer Herbert Saffir, an expert on designing buildings with high wind resistance, to create the Saffir-Simpson scale, which rates the strength of hurricanes. Back in 1971, when minimalism was in fashion, their categories ran from a simple one to five.[1]

Because tornadoes are different kettles of fish, Ted Fujita, originally a professor in Japan before coming to the University of Chicago in 1953, created a scale just for them—also in 1971. The original Fujita scale had thirteen levels, F0 through F12, the highest levels of which were purely theoretical—imaginary winds raging at the speed of sound.

But he had created too many categories. Fujita saw the light and chopped the number of varieties down to six. These days the scale, which was further tweaked and renamed the Enhanced Fujita Scale, also takes into consideration the degree of expected damage. The weakest tornadoes are thus EF0 and EF1; the strongest is EF5.

Of even greater importance, Ted Fujita, who died in 1998 at the age of seventy-eight, also discovered the microbursts and downbursts that destructively tumble from the bottom of thunderstorms. These may actually be more interesting phenomena than tornadoes, simply because most of us occasionally encounter them firsthand.

Thunderstorms are wind machines. Here is animated air

motion made darkly visible. They're even easy to understand. You don't have to be a Galileo to figure out what's happening. You start with a hot summer day, the sun heating the ground, which warms the air just above it. Warm air rises, so up goes a gas bubble like a hot-air balloon. This is called convection. It's invisible, although planes flying through such rising air get very distinct bumps of turbulence.

As we've seen, temperature normally falls rapidly with altitude. So a rising surface-heated air glob is warmer and thus lighter than the surrounding cool air and thus keeps ascending until it cools to achieve equilibrium with its surroundings. But if the day is humid, the rising air package remains *much* "lighter" than the air around it, so it keeps going up, sometimes to near-stratospheric heights. Eventually it cools to its dew point and can no longer hold its

Wind normally increases with altitude. At the 20,000-foot summit of Mount McKinley, left rear, winds howled at forty-five miles per hour, a common value, blowing snow and creating a standing wave lenticular cloud near the author's chartered plane, in 2014. *(Anjali Bermain)*

moisture, which suddenly condenses into countless billions of teensy droplets. A cloud is born.[2]

Hot air from below keeps rising to feed this cloud. It pushes sections of the cloud higher, forming a menacing cauliflower shape that can top forty-five thousand feet, beyond what any airliner can reach. Meanwhile its droplets rub together to create static electricity. At the same time, since you can't have a vacuum on the ground below, surrounding air is pulled in. The air pageant is now getting more and more animated. When rain starts forming and falling, cooling the air within the cloud, this denser cold air plummets, intertwined with the rain.

Now you have scary-strong winds. Some warm air is still rising into this "mature" thunderstorm while adjacent streams of cool air are plunging down from it. Even in a moderate storm the downdrafts register twenty-two miles per hour to match the speed of the surrounding downpour. If you're in a small plane you suddenly find yourself pushed earthward like a giant metal raindrop. You aim your nose upward and add all the power you can and hope you can outclimb it.

The downdraft, which can be a half mile wide, now hits the ground and, unlike the liquid rain, spreads radially outward in all directions, bending trees horizontally and inverting umbrellas. The violence contains complex turbulence because nearby updrafts can rise at this same speed. Now your plane, having successfully passed through the downdraft, encounters an adjacent column of fast-rising air. Suddenly you're sucked up toward the angry black cloud above. You push forward on the yoke, pointing your nose down. In many documented cases, the pilot wasn't able to dive fast enough to counterbalance the wildly rising air, and the plane was pulled into the roiling cloud as if by a conscious, malevolent monster.

No wonder all aircraft give thunderstorms a wide berth. Once, near Hartford, Connecticut, I experienced a frightening rapid downdraft when flying in clear skies, fifteen miles from where a storm had passed a half hour earlier. The air was *still* plunging violently.[3]

Such intensity turns up the juices, but real "fun with wind" lies in everyday life. Sure, it's cool to have some background, to know that temperature differences give birth to air motion. And that the shape of the terrain—for example, a narrow valley that's aligned with wind direction—can funnel it to faster speeds. Still, the act of simply observing the wind as it momentarily alters the world can provide immense enjoyment.

Everything depends on speed. Most nature-challenged people think in broad terms of a day as being "not windy" or "windy" or "very windy." What Beaufort brought to the table was a way of precisely translating wind's speed to what it does.

Francis Beaufort (1774–1857) had been shipwrecked as a teenager because of a poor navigational chart and consequently developed a lifelong obsession with making the seas safer through better maps and better understanding of the wind.

He started his seagoing career on a merchant ship belonging to the East India Company, then he joined the Royal Navy and worked his way up from the rank of midshipman to the rank of lieutenant during the Napoleonic Wars. He became a commander at the age of twenty-six.

He was badly injured in duty twice, but it never made him gun-shy or ocean-wary. His steadiness earned him growing admiration. He impressed everyone with his dedication, attention to detail, and scrupulous notes about sea conditions. He was the archetypical meticulous British commander, an inspiration for one of the characters in Gilbert and Sullivan's *H.M.S. Pinafore:* "I am the very model of a modern Major-General."

He became a captain in the Royal Navy in 1810 and spent his leisure time measuring shorelines and making improvements to charts. His reputation for intelligence, leadership, scientific integrity, and energetic devotion steadily grew and spread through the naval bureaucracy until even members of the nobility knew his name.

He accepted invitations to join the Royal Society and the Royal Observatory, helped found the Royal Geographical Society, and met all the great scientists of his day. As a top administrator, he helped coordinate the work of Britain's geographers, astronomers, oceanographers, and mapmakers and arranged funding for scientific expeditions. Beaufort trained Admiral Robert FitzRoy, who was appointed to command the survey ship HMS *Beagle* for what would become her famous second voyage. He recommended that "a well-educated and scientific gentleman" named Charles Darwin be invited as the captain's companion. As we all know, Darwin used this voyage's discoveries to create his theory of evolution, presented in his book *On the Origin of Species*.

So Beaufort was no mere hobbyist with a home-built anemometer. In 1805 he drew on his own wind observations and those of others, especially Daniel Defoe, who later became famous for his novel *Robinson Crusoe,* to create the air-motion scale that has borne his name ever since.

(Mark Twain said, "Everyone is a moon, and has a dark side that he never shows to anyone." It didn't seem as if Franicis Beaufort had any unseen sides to his celebrated meticulous personality. But after his death in 1857, his private letters were assembled, and many were found to be written in a personal code of his own design. It was a good code for its time, but it was readily deciphered by experts and found to reveal numerous confidences about personal problems, conflicts with colleagues, and secrets of a sexual nature. The moral might be: shred your documents if you don't want your secrets published posthumously.)

By the late 1830s, the Beaufort scale was made standard for ship's log entries on Royal Navy vessels. By the 1850s, others had adapted it to correspond with anemometer readings so it could be used on land. Then in 1916, as steam power made sailing ships obsolete, Beaufort's descriptions were changed so that they described how the ocean, not the sails, behaved. Further additions improved the scale's validity for land-based observations.

It has recently, of course, fallen out of popular use. In modern times, one never hears anyone say, "Look outside: there's a strong breeze of Beaufort force six." And on the rare occasions when Beaufort's highest number dramatically envelops your neighborhood in eolian violence, you'd think "hurricane" rather than "Beaufort force twelve."

I've nonetheless heard those exact words at sea. In 2006, when I was the astronomy lecturer on a month-long Holland America cruise around South America, our ship hit crazy winds as we approached Tierra del Fuego, at the bottom of Chile. We were inside the famous roaring forties, the name given to the routinely strong westerlies between latitude 40° south and latitude 49° south. Although no Pacific storm had been forecast, the winds just kept growing and growing until the captain made the announcement, "The winds are now Beaufort force twelve."

The ship rose and fell crazily. The piano slid across the library and smashed into the wall. Dishes kept crashing. Staying in one's cabin provided little relief: you slid down your bed from the disconcertingly steep seesaw bow-to-stern tiltings. I have it all on video. I went up on deck, which was totally empty except for the rare crewman who stepped out to gawk. Several told me that they'd never before been in a hurricane at sea until then. The waves looked just like Francis Beaufort's descriptions. Swells reached up to my eye level, seven stories above the sea. These were seventy-foot waves.

Yet everyday winds can be just as exciting when you're attuned to them. Here are the Beaufort numbers, along with the more useful miles-per-hour equivalents and—most important—how you can tell what's what.

Beaufort force 0 means no wind. The official description is "calm." Smoke rises vertically. Water is mirrorlike. Foggy nights are often like that.

Beaufort force 1 is officially termed "light air." The speed is 1–3 mph (1–5 km / hr). Rising smoke drifts. Weather vanes still don't budge. Small ripples appear on water surfaces, but these ripples have no crests.

Beaufort force 2, officially termed a "light breeze," has a speed of 4–7 mph (5.6–11 km / hr). Now you readily feel the wind on your skin. Leaves rustle, but no branches move. Weather vanes begin to turn. Small wavelets develop, but their crests are glassy, not rough.

Beaufort force 3 is a "gentle breeze," with winds of 8–12 mph (12–18 km / hr). Leaves and small twigs move constantly. Lightweight flags extend. Large wavelets appear, with breaking crests and a rare whitecap.

Beaufort force 4 is a "moderate breeze," with winds of 13–17 mph (20–28 km / hr). Dust and loose paper is lifted. Small branches begin to move. Small waves appear with many whitecaps.

Beaufort force 5 is a "fresh breeze," with winds of 18–24 mph (29–38 km / hr). Medium-size branches move. Entire small trees sway if they're in leaf. Most waves have whitecaps, and there is some spray.

Beaufort force 6 is a "strong breeze." Winds are blowing around 25–30 mph (39–49 km / hr). Large tree branches move. Telephone and overhead electric wires begin to whistle. Umbrellas are difficult to keep under control. Empty plastic garbage bins tip over. Large waves form; whitecaps and spray are prevalent.

Beaufort force 7 is a "moderate gale." Winds blow at 31–38 mps (50–61 km / hr). Large trees sway. Walking requires some effort. The sea heaps up. Some foam from breaking waves is blown into streaks along the wind direction.

Beaufort force 8 is a "gale." Winds are 39–46 mph (62–74 km / hr). Twigs and small branches get broken from trees and litter the roads. Walking is difficult. Moderately large waves form, with blowing foam.

Beaufort force 9 is a "strong gale," with winds of 47–54 mph (75–88 km / hr). Some large branches snap off trees. Large trees sway wildly. Temporary signs and barricades blow over. High, twenty-foot waves produce rolling seas and dense foam that reduces visibility.

Beaufort force 10 is a "storm" or "whole gale." Winds roar at 55–63 mph (89–102 km / hr). Weak trees are blown down or uprooted. Saplings are bent and deformed. Weak or old asphalt shingles are peeled off roofs. Large waves of 20–30 feet have overhanging crests. There is a heavy rolling of the sea, which is white with foam. Visibility is reduced.

Beaufort force 11 is a "violent storm," with winds of 64–72 mph (103–117 km / hr). There is widespread damage to trees and crops. Many trees are blown over. Many roofs are damaged. Many objects left unsecured are blown away and can break glass. The sea has very high waves of 37–52 feet and extensive foam, and there is restricted visibility.

Beaufort force 11 winds of around seventy miles per hour are slightly below hurricane strength but have easily destroyed half the trees in this forest.

Beaufort force 12 is a "hurricane." Winds are above 72 mph (above 118 km / hr). Crops, plants, and trees suffer widespread damage. Some windows may break; mobile homes and weak sheds and barns are damaged. Heavy debris is hurled about. Waves are huge; over 50 feet. The sea is completely white with foam and spray. Visibility is negligible, thanks to driving spray.

I'm offering the entire Beaufort scale for one reason only: because to recognize what is happening and be able to label it means that you can watch air motion with more attention. And better observation creates better enjoyment. This way if you see branches swaying yet large tree trunks are steady, and overhead wires are whistling, and your plastic garbage pail just blew over, you can say with confidence: "Hey, honey, there's a *strong breeze* outside. The wind is between twenty-five and thirty miles per hour." And you'll earn a perfunctory nod from your spouse, who really doesn't care.

No matter. You and your fellow nature lovers are enraptured by the magic of the wind. And perhaps reminded of Alhazen, who figured out a thousand years ago where it ends. And of Evangelista Torricelli, whose words still linger:

"We live submerged at the bottom of an ocean of...air."

CHAPTER 10: *Falling*

Enigmas of the Most Far-Reaching Force

The whole damn thing, the universe,
Must one day fall.
— HOWARD NEMEROV, "COSMIC COMICS" (1975) .

We take for granted the tinkle of falling rain. And we stick to Earth's surface without giving it a thought. Yet can anyone honestly explain gravity or claim to know what's going on? The otherwise perspicacious cultures of the ancient Greeks, Chinese, and Mayans didn't even try.

Even today, how many of us ever really pay attention to the act of falling? Any kid who has belly flopped into a swimming hole knows that the higher the dive, the more it hurts. That's because the farther up we jump from, the faster we hit the water. In fifth grade they used to cite thirty-two feet per second per second as the rate at which a plummeting body accelerates, until primary school science switched to the metric system, and then it became 9.8 meters squared. Too bad. We might have paid attention if they'd expressed it in everyday language.

Fall for a single second and you hit the ground going twenty-two miles an hour. Simple.

Each additional second you're airborne makes you land another twenty-two miles an hour faster. Still simple.

If you want to stay in the air for exactly one second, you have to jump from a height of sixteen feet. One and a half stories. If you

141

land on a trampoline this might not hurt. But you should not, as they say on TV, try this at home. To stay aloft for two seconds, however, means leaping from a six-story roof, and you'd then accelerate to forty-four miles an hour, the impact from which is usually not survivable. So humans, unlike squirrels, have a very narrow available range of safe falling. One second of plummeting can sometimes be okay. Two seconds means death.[1]

This is the hard-nosed reality all people and animals confront from the moment they take their first baby steps. Our motions are a contest between our muscles and the ground beneath us, as it eternally holds us as closely as possible.

As we've seen, Aristotle and his friends tackled downward motion by saying that everything made of the elements water and earth wants to fall. In a straight line. After all, a stone tossed off a cliff angles more and more toward a linear trajectory with every second it keeps plummeting.

Watching the sky, the ancients decided that up in the heavens, objects want to move in circles. The sun and moon daily circle around us. The stars nightly wheel around the North Star.[2]

Moreover, the only celestial objects that aren't dots are the disks of the sun and moon—more circles. Clearly, the gods like circles in their realm. You can't blame them. The circle, according to the Greeks, is the perfect geometric form—so perfect it was divine, with a legacy that lingers to this day in traditions such as the exchange of rings at weddings and engagements. It's the only shape whose boundary contains no special points or direction changes and whose edge is at every point equidistant from the center.

So according to the Greeks, all motion is either linear or circular. Circular up there, linear down here. There was no word for gravity. There wasn't even the concept of a downward-pulling force. Instead, objects themselves "want" to head downward and will do so the moment obstacles are removed.

142

That's how things stood for century after century while children tripped and scraped knees and old men idly threw pebbles into ponds.

It took until modern times for this gravity business to become central to space exploration and bungee jumping and such. Meanwhile the parallel topic of air resistance developed into a major study of its own. It's a central tenet of aircraft engineering and parachuting and was always a design feature in the animal kingdom—which explains why cats and squirrels usually never accelerate to lethal speeds no matter how far they fall.[3]

All that technological fun stuff was still to come when the ancient Greeks were alive. But when the sixteenth century dawned, science was dealing with old Greek views that had since been incorporated into Church dogma and was trying to make those square Plato-idea pegs fit into the round holes of actual planet movement.

The problem was that the holes were *not* round. They were oval. To chart planetary motion against the background stars was to observe loopy trajectories that didn't jibe with anything circling a stationary Earth. Then in the sixteenth century, nightly observations performed meticulously for twenty years by the obsessive Danish astronomer Tycho Brahe refined the length of the year to within an accuracy of *one second*, which proves he was a type-A fanatic who refused to "round off" anything. He was *good*. Yet he failed to unlock the simplest underlying secrets of celestial choreography.

It did not appear as if planets moved in circular paths. But Tycho assumed all sorts of comically jury-rigged systems—planets moving in circles around empty spots of space that in turn orbited still more circles that then circled us—to preserve the traditional Holy Roundness. And to keep a motionless Earth at the center of it all. All his mental gymnastics, a pathetic purgatory of years of intellectual labor, served the single desperate purpose of trying to make the universe jibe with the clergy's mistaken view of natural motion.

When Tycho died in 1601, his assistant, Johannes Kepler, inherited his notes and pondered them for the next ten years, right through and beyond Galileo's first telescope discoveries in 1610. Kepler, a brilliant mathematician, came to a startling conclusion. The celestial minuet made sense only if the sun lay at the center of all motion, and the planets — including Earth — moved in *elliptical* paths.

Ellipses were not sexy, then or now. But they are the fact of the cosmos. They are the way gravity makes nearly all celestial bodies move.

Understanding ellipses is easy if you draw one. Push two thumbtacks partway into a piece of plywood or cardboard and loosely put a loop of string around them. Insert a pencil within the loop, pull it firmly sideways, and you'll draw an ellipse. In the actual universe, each of those tacks is called a focus, and the sun occupies one focus of every planet's orbit. (The other is just an empty spot of space. This bothers some people, who feel that such a vital mathematical point merits more than mere vacuity. Well, perhaps some enterprising space-travel company will someday erect a floating café there with a catchy name like the Focal Point.) This is the simple and complete reality of every planet's path through space.

Kepler found that each planet speeds up as it approaches the sun in its oval path but decelerates as it heads away. Holy cow: Earth and all other worlds *continually change their speeds*. Nobody had seen that coming.

Obviously, something about the sun pulls on the planets. It was a puzzle pondered at that same time — the opening decade of the seventeenth century — by Galileo Galilei.

Galileo, who, like Kepler, was a longtime subscriber to *Heliocentrism Today*, decided to study the way things move and fall. He built ramps that had various kinds of slopes, set balls rolling, and watched what happened. He timed things carefully and concluded that, regardless of how steep or gentle the slopes were or how high a ball was when it was released, it would speed down one incline and then up another *until it reached the same height as the one from which it was dropped.*

If the second ramp was perfectly flat—horizontal—the rolling ball would just keep going and finally stop only, Galileo correctly determined, because of friction. He was struck by an amazing thought. Maybe the moon and planets were rolling sideways, too. In which case they'd continue in motion forever, which is exactly what they appear to be doing.

He used simple math, and the numbers added up as long as planets were not slowed down by any air resistance. They must be orbiting in a realm of emptiness!

These days we're accustomed to the idea of space being a vacuum. But back then, "nothing" had a long, bumpy history in philosophy and never came out smelling like a rose. The Greeks, for example, made many intriguing arguments as to why nothingness was impossible, while Renaissance clergymen reasoned that "God is everywhere, so there can be no vacuum."[4]

Galileo, in the opening years of the seventeenth century, became the very first person who was sure he knew what existed "up there" in the heavens. Nothing.

In his heretical publication *The Starry Messenger,* Galileo didn't make much ado about nothingness only because it was the least of his revelations. He also boldly claimed that Aristotle was wrong when he said that massive objects fall faster than skimpy ones. After all, Galileo's biggest metal balls rolled no faster than his lighter models. Instead, he proclaimed, it was merely air resistance that slowed down spread-out objects, such as feathers. (We got to see Galileo's breakthrough notions come true on the moon when astronaut David Scott simultaneously dropped a hammer and feather and both plummeted in perfect sync. He did this near the end of the Apollo 15 mission in 1971.)

And now we come, inevitably and predictably, to Isaac Newton. Who, by the way, indeed told at least four people that he got his inspiration about gravity from watching a falling apple. The only false aspect of his popularized, *American Idol*–like bio is the business about a Golden Delicious bonking him on the head.

Pondering the way the moon and apples behave, Newton realized that Galileo had been onto something. Both objects move the same way. Abandoning the Greeks' longstanding linear-versus-circular reasoning, he unified the heavens and the earth by coining a new word: *gravity*. He created it from *gravitas*, the Latin word for "heaviness." What exactly it was he could not say. But how it acted—ah, this he could quantify perfectly.

Actually, his contemporaries, including Robert Hooke and Edmond Halley, also assumed that some mysterious force pulled objects toward the center of the earth. Halley even presumed it grew less powerful with distance and, in a now-forgotten experiment, carried a pendulum to the top of a 2,500-foot hill, where, he claimed, he watched it swing a bit more slowly there. These natural philosophers, as scientists were then called, not only believed the planets were yanked by the sun but also correctly said that this force grew proportionally weaker with the square of their distance. Meaning an object that's three times farther away from the sun than you are experiences three times three, or nine, times less of a pull.[5]

So Newton hardly came up with the idea of the force, even though he named it and introduced it to the Western world. Rather, he accurately described how it behaved.

Isaac Newton was born in 1643, in Lincolnshire, England, one year after Galileo died. He studied at Trinity College in Cambridge and later became professor of mathematics there. In a paper published in 1687 that soon became widely known as the *Principia*, Newton mathematically proved that the sun's gravity should make planets travel in elliptical paths, thus effectively awarding Kepler a posthumous 1600 SAT score. It was here that he offered his famous three laws of motion, although in fairness Galileo had already pretty much stated the first two:

1. Every body continues in a state of rest, or of uniform motion in a straight line, unless it is impelled to change that state by forces impressed upon it.
2. The change of motion is proportional to the force impressed, and it is made in the direction of the straight line in which that force acts.
3. To every action there is an equal and opposite reaction.

In plain language, a moving body tends to keep moving. And a stationary body likes to remain at rest. Both of these tendencies are called *inertia*. Newton also introduced the concept of *momentum*. Momentum involves exactly two things: the mass of an object (what one perceives as its weight) multiplied by its speed. A slow-moving truck may move at the same speed as a bicycle, but the truck has more mass and hence more momentum. It's harder to stop.

Newton was also the first to state the obvious — that the strength of a force is determined by how it influences a body's motion. He also spoke of acceleration as a change in movement, whether of speed or of direction.

Newton regarded gravity as a force simply because it changes the way objects move. It pulls them ever faster. Here on Earth, we know that gravity yanks things toward Earth's center at the rate of twenty-two miles per hour faster every second. Newton's brilliance was in understanding that a falling apple behaves exactly the same way as the moon orbiting around us does.[6]

Newton's third law expressed something totally new: the notion of equal and opposite reactions. This means that any object exerting a force also feels it acting on itself. If you push a stalled car, you feel that same force upon your hands, pushing back at you.

The exploding charge propelling a bullet forward also creates a recoil in the rifle. Because the bullet weighs less than the rifle, one object enjoys more forward speed while the heavier one moves backward with less oomph. This inequality business goes off the scale when it comes to events that involve our planet. If you jump up, you are simultaneously pushing Earth backward in the opposite direction. However, since Earth weighs an octillion times more than you, it moves an octillion times less than you do when you jump.

That equal-but-opposite law showed why a rocket sending a stream of high-speed gas out its bottom end, even in the vacuum of space, moves the opposite way—upward. There's no need for those gases to push against anything.

Using Newton's figures and just a tiny bit of math we can figure out how fast a bar of bullion would fall if tossed exuberantly from the Fort Knox roof by a gold-standard extremist. Or how fast a midlife-crisis bungee jumper falls when leaping from a twenty-story bridge. Grab a one-dollar calculator. Don't be afraid. It's time for the "fun with math" segment.

Multiply the jumper's altitude (in feet) by 64.4, then hit the square root button. That's his final speed in feet per second. If you prefer miles per hour, multiply this by 0.68.

Let's consider one example. Our bungee jumper leaps from two hundred feet, so this figure times 64.4 is 12,880. The square root of this is 113. And that's his final speed: 113 feet per second. Multiplying that by 0.68 yields seventy-seven miles per hour. Not so difficult.

The greatest speed you'd reach if you jumped from the highest possible perch—by falling toward Earth from even beyond the moon, say—would be 25,031 miles an hour, ignoring air resistance. It's exactly the same speed needed to *escape* from Earth with a single upward blast, as though you were a circus performer fired from a cannon. So the velocity needed to escape any celestial body is also the velocity at which you'd land if you fell there from a great height.

This up-speed-equals-down-speed business is pretty cool. Toss an orange up and let it land back in your hand. Interestingly, the exact speed at which you chose to toss it up is the same speed at which it's coming down when you catch it.

Each celestial body has its own impact speed, or escape speed, predetermined by its mass and its diameter. For the moon that's 5,368 miles an hour. For the sun it's more than a million miles an hour, or 384 miles per second. That's how fast a drifting, out-of-fuel spacecraft piloted by incompetent aliens would be pulled into the sun by its gravity.

On Earth, air resistance slows things down. In skydiving class they make you practice spreading your arms and legs to let your body present its maximum surface area to the wind. If you do that you won't gain any additional speed beyond 120 miles an hour. This is the famous "terminal velocity."[7]

It's reached pretty quickly, after jumping from a height of just five hundred feet, or fifty stories. So, perhaps surprisingly, you'll not go any faster if you jump from the 110th floor than you would if you jump from the fiftieth floor. Daredevils who bypass the fiftieth floor in favor of leaping from the roof much higher up merely want to buy themselves extra air time for their parachutes to open—an excellent idea.[8]

But we still haven't explained why all this happens. So let's fast-forward to Albert Einstein, born in 1879.

In his 1905 and, especially, his 1915 relativity theories, Albert Einstein did not just tweak Newtonian mechanics. He tossed it out, replacing it with concepts so bizarre that even now, a century later, they remain mind-twisting. It was a brand-new way of thinking about movement in the universe.

Einstein would not have invented a better mousetrap if the old one worked just fine. But the behavior of celestial bodies contained a few slight but inexplicable wrinkles when examined through the lens of the old, simple, Newtonian calculations of force and mass and acceleration.[9]

Einstein decided that gravity wasn't a force at all. In a leap of inspiration unequaled before or since, except perhaps among the quantum gang of Heisenberg and company, Einstein said that an unseen matrix he called *spacetime* pervades every cosmic nook and cranny. An amalgam of time and space, its configuration dictates how any object must move through it. An object's very presence, its mass, distorts the surrounding spacetime. Anything moving through this region has its trajectory of motion, as well as its passage of time, change in a predictable way.

By this thinking, the sun doesn't pull on our world. Instead, Earth merely falls in the straightest, laziest, most direct path through the local curved spacetime. Our nearby sun's enormous mass depresses spacetime like a heavy ball resting on a rubber sheet and making it sag. Earth rolls along this warped rubber membrane and curvingly arcs back to its starting point after a year.

Nor is spacetime limited to faraway places. It's also right here in the room. We stand on Earth's surface and feel the ground pushing up against our soles and heels. That's because we experience Earth's motion and ours through the local spacetime, which has been distorted by Earth's mass.

So Einstein replaced gravity with geometry. Every object's path is dictated by the configuration of the local spacetime. As a close-up example of how it works, consider two batters stepping to the plate. The first hits a pop-up that travels skyward a great distance and stays aloft a long time before it's caught by the shortstop. The next batter hits a line drive. It takes a more linear path before being snagged by the same shortstop and gets there much faster.

To our minds, which regard time and space separately, these two hit balls take very different trajectories. They appear to be dissimilar events. But plotted in the single matrix of spacetime, they took identical paths. Indeed, whenever objects are released to travel on their own (as long as they leave from and arrive at the same two points) they must follow identical geodesics (paths through spacetime). Only to our human perceptions does each consume a different amount of time and a dissimilar route through space. In truth, the two are so linked that should you alter the time path of an object (e.g., make the ball stay *longer* in the air) it automatically changes the space path.

Unfortunately, Einstein's field equations for the way spacetime is warped and the way objects move through it are incredibly complex.[10] They're so labor intensive that even NASA doesn't use them when they calculate spacecraft travel routes to the planets. They prefer to stick with Newton's simpler math, which yields results that are good enough and far easier to manage.

Today's schoolchildren are still usually taught the older, Newtonian viewpoint, that Earth circles the sun because of solar gravity. Few science curricula provide children with the superior Einstein concept, that our planet merely falls along a straight path (geodesic) through the curved spacetime produced by the nearby massive sun.

We could end the story of dropped keys and whizzing planets right here, except for one problem. Whether we call it distorted spacetime or gravity, the phenomenon of objects being pulled toward others remains mysterious. After all, spacetime is a mathematical model, not an actual entity such as Swiss cheese. Time has no independent existence on its own except as a way we humans perceive change. Space, too, is not a real commodity. We cannot bring it to a lab and analyze it as we would a piece of quartz. Space-time is an accurate mathematical way to describe and predict motion; it is not an ultimate explanation. Many physicists still prefer to speak of gravity as if Einstein never existed.

We may someday find out why objects pull toward other objects. If gravity is a force, there ought to be a force-carrying particle that brings it from one place to another. After all, photons (bits of light) are the force-carrying particles that transport electromagnetism. Einstein postulated "gravitons" as gravity's butlers. So far, however, they have not been detected. (Although if gravity is nothing but a kind of geometry, a distortion of spacetime, then perhaps force carriers may not be needed.[11])

Does gravity's power depend on the rest of the universe? Does it somehow involve hypothetical strings? Does the gravitational "constant" change as the universe expands? Will Earth's gravity grow weaker over time? Can gravity be some sort of influence from another dimension?

Gravity's enigmas, like autumn's falling apples, Newton's original inspiration, still plop all around us.

CHAPTER 11: *Rush Hour for Every Body*

Revelations Gained by Looking Within

And the heart must pause to breathe...
—LORD BYRON, "SO WE'LL GO NO MORE A ROVING" (1830)

Sorry, I'm busy right now," you tell a friend.

That's so true. Your body is as busy as the galaxy.

Even when we're resting and daydreaming, internal activity is nonstop. Some of it is obvious. We can feel our pulse, our heartbeat, our heaving chest. Maybe our stomach gurgles for a moment. Not much else. This limited awareness of a mere handful of internal motions is a good thing. Nature has spared us from being overwhelmed by its myriad under-the-skin dramas.

But let's be aware now. If only to appreciate the exquisite, epic complexity involved when a teenager applies eyeliner.

We might start by thinking about thinking. The brain, of course, is the crown jewel of our nervous system. (Or is the brain just blowing its own horn at this moment by making me write this?) It has eighty-five billion neuron cells and, even more impressive, boasts 150 *trillion* synapses. These are its electrical connections, its possibilities. This figure is nearly a thousand times greater than the number of stars in the Milky Way galaxy.

The number of brain neurons is staggering. To count them at the rate of one per second would require 3,200 years. But the number of brain synapses, or electrical connections, is beyond belief.

Those 150 trillion could be counted only in three *million* years. And that's still not the end of the matter. What's relevant is how many ways each cell can connect with the others. For this we must use factorials. They're very cool. Let's say we want to know how many ways we can arrange four books on a shelf. It's easy: you find the possibilities by multiplying $4 \times 3 \times 2$, which is pronounced "four factorial" and written as 4!—i.e., twenty-four. But what if you have ten books? Easy again: it's 10!, or $10 \times 9 \times 8 \times 7 \times 6 \times 5 \times 4 \times 3 \times 2$, which is—ready?—3,628,800 different ways. Imagine: going from four items to ten increases the possible arrangements from twenty-four to 3.6 million!

Bottom line: possibilities are always wildly, insanely greater than the number of things around us. If each neuron or brain cell could connect with any other in your skull, the number of combinations would be 85 billion! (i.e., eighty-five billion factorial). This is more zeros than would fit in all the books on Earth. And that's just the zeros—the mere representation of the number, not the actual number. Remember, each time you add just six more zeroes, you've made it express a quantity a million times larger than everything that went before that point. The brain's connection possibilities lie beyond that same brain's ability to comprehend it.

All this architectural complexity lies in what may seem like an inert three-pound lump of cheese about the same size as a piston on a 1400 cc engine. Because there are no muscles in the skull, and because the brain has barely more density than water, it does indeed appear to be a mushy, unimpressive lump. But its animation is utterly disguised. What makes it vibrant are its relentless electrical activities. Unseen sparks fly everywhere. Each neuron functions on about one hundred millivolts. A tenth of a volt is darned efficient: this operating matrix is less than that of an AAA battery. Even if you add up the brain's entire energy consumption, it's a mere twenty-three watts (for a person consuming 2,400 calories daily).

Still, the brain uses a whopping 20 percent of the body's energy despite taking up only 2 percent of the body's mass. It's an energy hog. There's no off switch; the current courses continuously.

The first hints of this electrical activity came from Luigi Galvani in 1791, when he published his work on the electrical stimulation of nerves in a frog. If electricity makes muscles contract, then that's how the brain must accomplish its commands! The very next year, a fellow Italian, Giovanni Valentino Mattia Fabbroni, suggested that such electrical nerve action must employ chemicals. The whole idea got a big boost eight years later, in 1800, when Alessandro Volta invented the wet cell battery. Here was electricity generated and contained in a self-enclosed way—why couldn't it be the same in the brain?

Of course it was far more complicated than that. When the 1906 Nobel Prize in Physiology or Medicine was awarded to Camillo Golgi and Santiago Ramón y Cajal for their breakthroughs on the organization of the nervous system, it merely marked an early step in probing a labyrinthine structure that even today is far more mysterious than any other part of the body. But at least at that point we grasped the mechanism by which muscles are commanded to move.

Electricity through a copper wire travels at 96 percent of the speed of light. No such luck when it comes to neural strands. Our body's neurons come in several different varieties and capacities, but none lets current flow even 1 percent as swiftly as it does through an electric can opener. Yet we apparently don't need such light-speed cognition to accomplish everyday mental brilliancies, such as bagging the garbage. Our actual maximum operating rate of just 390 feet per second, or less than a millionth of the speed of light, is fast enough to do the job.

This becomes obvious with a quick experiment. Close your eyes and flail a hand rapidly around—over your head, to the sides, anything. You're always aware of exactly where it's located, every moment, no matter how quickly you alter its position. Your in-the-moment cognizance of your hand's location proves that neural electrical signals reach your brain extremely quickly, since only "real-time" information would be useful in such situations. In fact, those impulses travel at 250 miles an hour.

That's the nerve transmission speed for essential stuff. But what qualifies as "essential"? Fortunately you don't have to prioritize the relative importance of all the sensory, muscular, pressure, pain, and other signals the brain receives. It's taken care of, designed and hardwired before you even left the womb. A friend's carelessly exuberant hand gesture is about to poke your eye? You instantaneously blink and evade. You're eating and would prefer not to stab yourself with the fork? The positional signals from your fingers and lips coordinate in the moment. On an overnight camping trip, stepping out of the tent barefoot, you tread on a suspicious object that feels an awful lot like a snake. You yank your leg up in an eyeblink. All these reflexes were neurally commanded at 250 miles an hour.

But now stub your toe. Or just remember when you did. It took several seconds to feel any pain. That's because pain signals travel along separate cables at a low-priority speed of just three miles an hour, or two feet per second. There's no rush to deliver bad news.

How about thinking? These signals occur at an in-between speed. Neither fastest nor slowest. They slither and branch through the cerebral cortex at seventy miles an hour. The process is speedy enough so that you can make decisions before a separate circuitry— your ego, your sense of yourself—is informed of their completion.

In 2006, the late researcher Benjamin Libet and his team instructed volunteers to push a button the moment they decided which arm they intended to raise and then immediately lift the appropriate arm. Here's the amazing thing. Researchers watching the subjects' brain waves could reliably tell which decision they had made up to ten seconds before the volunteers themselves were aware of their choices!

In other words, the brain's electrical activity performs automatically, like the functions of the pancreas or liver. It makes decisions autonomously. Only a bit later do we realize what's been decided.

We may have a subjective awareness of choosing. We may say, "I decided to have Chinese food tonight instead of Italian." But we actually exercised no free will at all. The brain decided on its own through spontaneous electrical connections. None of us has the slightest idea how to control this activity any more than we can manipulate the workings of our kidneys. (If, disagreeing, you now claim that you *can* personally make choices and that decisions are *not* automatic, you should know that the construction of that very thought happened autonomously before you even intended to think it or say it.)

All these fast, slow, and intermediate electrical impulses and synaptic connections happen continuously, and their pace is fastest in the morning. We get a break only when the lights go out: the brain operates at a much-reduced level when we're asleep.

The nervous system's activity, which peaks between the ages of twenty-two and twenty-seven and starts to diminish thereafter, is of course the control system for myriad other internal motions. The ones we're most aware of, of course, are the breath and the heartbeat.

Figuring out even basic heart realities didn't come easily. Despite the knowledge gleaned from dissecting cadavers (during the centuries when this was ethically acceptable), people found the function of the heart and its drumbeat bewilderingly mysterious until rather recently. As long ago as 4 BCE, Greek physicians were aware of heart valves and arteries but still came to wrong conclusions about them. Because blood pools in veins and not arteries after death, Greek anatomists wrongly assumed that arteries transport air throughout the body. Erasistratus, a physician in Alexandria who died in 250 BCE, said that when people received cuts to arteries and bled, it was merely because those "air-filled vessels" were suddenly flooded with blood from the veins.

Later, the famed Greek physician Galen, in the second century CE, did maintain that both arteries and veins contain blood. But he did not think it was the heart that pumped it. Rather, he said, the pulsing arteries did the pumping. The heart merely sucked in blood and served as a kind of repository. Nothing circulated. Blood was created by the liver and then somehow got used up and continually replaced by new stuff.

It wasn't until 1628 that physician William Harvey finally figured out the circulatory system and explained the reason for that thumping in our chests. (In keeping with science's hallowed tradition toward pioneers, Harvey was ridiculed for decades.)

The heart beats 2.5 billion times in a lifetime. The five quarts of blood an adult male continuously pumps (four quarts for women) flow at an average speed of three to four miles per hour—walking speed. That's fast enough for a drug injected into an arm to reach the brain in only a few seconds. But this blood speed is just an average. It starts out by rushing through the aorta at an impressive fifteen inches a second then slows to different rates in various parts of the body.

Normally, liquids such as water speed up when forced to flow through a narrow pipe. Kids like to squeeze a hose to make the water jump farther, to douse their friends. But the opposite happens in the narrow capillaries. Here is where blood flow is *slowest*.

It's all part of the oxygen-exchange plan. The reason goes beyond the fact that capillaries are farthest from the heart. Rather, there are so many of them that their cross-sectional area is greater than what's found in veins and arteries. The blood volume is essentially spread out there.

Lymph fluid moves, too, through its own system of channels, at a low speed of a quarter inch a minute. But air is much livelier. Men and women both normally inhale and exhale about a pint of air—half a quart—twelve or fifteen times a minute. This adds up to a gaseous intake just shy of two gallons a minute. To make this happen, the lungs and diaphragm move in and out an inch a second.

Meanwhile, in the always enjoyable food department, we pop a pastry into the mouth and chew, the lower teeth performing all the motion, rising and falling at the rate of an inch a second. (Studies show that more saliva squirts out when we're hungry.) Gulp, and down she goes, and now we rely on esophageal peristalsis, a wave of contractions that brings the food stomachward at the speed of three-quarters of an inch per second.

Splash—into the tummy. There it remains for an average of two to four hours.

Next the food is further processed, and its water content removed, as it chugs growlingly through the twenty-foot-long small intestine and then the six-foot-long large intestine. This putrefying mass barrels along at a speed that varies from a foot an hour to a foot every three hours. It depends on the person and also on the food. Stuff with a lot of roughage moves fastest. And fully half the weight of stool is bacteria. Indeed, research in 2012 revealed that about 3 percent of each of us is pure bacteria. We're each an "us" rather than an "I."

The entire process—in one end and out the other—can be over in a single day. Or it may require three days. There's no "normal" here, despite the fact that we all have an opinion about how often we should be going to the bathroom. Some individuals have bowel movements three times a day. Others just once every other day. If you want to speed things up, increasing dietary fiber to twenty-five or more grams a day is the best method. We cannot control the speed at which electricity travels through our neurons, lymph fluid flows through the lymphatic system, oxygen and carbon dioxide change places in the lungs, or blood flows through our capillaries. Nor can we alter the speed of asteroids. It is only in the personal digestive realm that we wish to become "control freaks" and obtain what we imagine to be optimum velocity.

Same with urination. Men and women big and small all pee at the identical average rate—between one-third to one-half ounce per second. Since the mean urine quantity is one to two quarts a day, we are condemned to spend one to two full minutes daily peeing. Rarely more than three. The average woman urinates eight times a day, the average man seven, though up to thirteen times a day is not, believe it or not, considered abnormal. To add it all up, a person who urinates seven times daily will require between nine and twenty-seven seconds to do the job per session.

We thus dedicate an entire month of our lives to this activity.

Men and women blink at the same rate, too. That is, we blink about ten times a minute, or once every six seconds. Staring—as we do when reading—cuts that rate in half. But while extended focusing on one visual task makes us blink less, being tired does the opposite and creates more blinking.

Infants blink just once or twice a minute, for reasons unknown. A possible explanation is that infants need less eye lubrication than adults do simply because their eyes are smaller. Moreover, infants do not produce any eye secretions during their first month of life and thus spare us the heartbreaking sight of crying with visible tears. Infants also get far more sleep than adults do. In any case, blinking steadily increases through childhood and matches the adult rate in adolescence.

Each blink takes just a tenth of a second. But their mysteries linger far longer. People with Parkinson's disease hardly blink at all, while schizophrenics blink more than people without that affliction. No one knows why.

The instigation of the eyeblink is even faster than the blink itself. The human eye's reflex, sometimes elicited by nothing more than a puff of air, takes between thirty and fifty milliseconds to travel from the brain to the eye. Better than one-twentieth of a second. Compared to that, voluntary reaction time to a visual signal in a laboratory, even when the subject is keyed up and expecting it, is around one-seventh of a second. In a car, it would then take an additional three-quarters of a second for a driver to move her foot from the gas pedal to the brake. Reflexive actions are the way to go. Conscious choices may be overrated.

Some body motions are even faster than eyeblinks. Within each cell, for example, protein synthesis creates new substances, each with a particular vital function. How fast? A cell's ribosome can make a disease-fighting protein in ten seconds. Given the millions of cells all simultaneously producing proteins to combat an infection, it's very long odds against invading bacteria ever gaining a foothold.

Good thing. These armies are often evenly balanced. A colony of bacteria can double its size in nine minutes and forty-eight seconds, and we've all experienced boils and other such germ "cities" that temporarily manage to get the upper hand against our defenses.

We're maximally exposed to disease pathogens when we travel, especially in crowded enclosures such as airliners or buses.

Which brings up the issue of our entire bodies in motion. Walking. All around us, people swing their limbs. The average person's leg and arm completes its back-and-forth cycle in about a second and a half. Meaning we take two full steps in three seconds.

We can deliberately choose to walk faster or slower, of course. Yet people and animals have a natural gait they'll unconsciously use whenever possible. A leg "wants" to oscillate just the way a playground swing does. It has a natural cycle whose period depends solely on its length. We all know that a playground swing on a nice long chain has a satisfying, long-period oscillation, while a short chain yields a quick, jumpy back-and-forth ride. This is the pendulum effect.

It was discovered by none other than Galileo. The place where

Galileo first noticed this fantastic property of naturally swinging objects is still there, unchanged to this day. If you ever go to the Leaning Tower of Pisa, you can't miss the duomo, the cathedral, adjacent to it. Within that huge, dark, musty space, chandeliers still hang on chains from the ceiling far above. Sometimes, especially on a windy day, these chandeliers display a slight, achingly slow, back-and-forth motion. In 1582, during mass, Galileo—perhaps so enraptured by the liturgy that he stared blankly up toward the ceiling—made the startling discovery that the length of time it took for these chandeliers to complete each swing never varied. It was about nine seconds, whether they were moving a few inches or several feet. This observation apparently stewed in his brain awhile.

The Piazza del Duomo, in Pisa, Italy, is the site of two great motion-related events. According to his pupil Vincenzo Viviani, Galileo dropped a light object and a heavy object from the leaning tower in 1589, proving that they fall at the same speed. Seven years earlier, in 1582, in the great duomo in the foreground, Galileo first noticed the pendulum effect.

A *long* while. Fully twenty years later, he decided to make a study of this pendulum effect. Beginning in 1602, he wrote to tell someone that the heaviness of the bob, or weight, at the end of a chain, wire, or string doesn't influence the period of the swing. Nor, essentially, does the amplitude. Meaning that if you give a child a tiny, barely noticeable push on a swing so that it only moves a few inches, the time it takes to go back and forth won't be different from what it'll be if you deliver a huge push that takes her high up, nearly sideways, and then wildly back up the other side.

Galileo realized that this property, called isochronism, would make pendulums useful as clocks. It means that the period of swing depends solely on the length of the string. There's a natural swing rate to everything. Any weight on a thirty-nine-inch string—very close to but not exactly one meter—creates a perfect one-second swing, meaning two seconds for the complete back-and-forth period. (This length might logically seem the basis for the meter, but no: the meter was originally fashioned to equal one ten-millionth the distance between the equator and either pole.) Eventually, grandfather clocks were duly built with pendulums of the correct length to *tick-tick* faultless seconds.

By chance, most men have legs of about that same length or a few inches more. As they swivel from their hip sockets, these hairy weights "want" to complete a back-and-forth swing in about 1.5 seconds.[1] For a five-feet-four-inch woman, the gait is a bit faster.

Again, the body naturally uses the easiest, least energy-consuming gaits. If you're in a hurry you can obviously expend extra energy and go as fast as you like. These days we physically hurl ourselves at speeds that would have bewildered all but the last eight of the 12,500 generations of Homo sapiens that went (more slowly) before us.

Do human travel velocities count as "natural" motion? We normally separate our technological triumphs from things like blowing sand and charging elephants. But maybe that's too arbitrary.

Our brains and our restlessness developed beyond our control; perhaps our own meanderings are as natural as that of the Mississippi.

So let's briefly outline how fast entire bodies move. This would
have been an easy task during the first two thousand centuries of
human history. We walked or ran. An hour of effort let us sweatily
advance ourselves between three and ten miles. After we tamed
horses, we galloped for short distances.

Average Americans walk 65,000 miles in their lives. More than
twice around the world. That's not so different from our ancestors.
But this is: we each *travel a million miles* over the course of a
lifetime. Such a degree of movement was unheard of until recently.
(And not just because the word *million* didn't exist until the four-
teenth century, before which the largest number was a myriad—
ten thousand.) Danger per mile was so much higher, even as
recently as the Civil War era, that few would have lived to accumulate that many frequent-traveler points. True, an extraordinary
nineteenth-century railway conductor or seaman might have
accrued enough to join the million-mile club—but he'd likely
have lots of scars to prove it.

The pivotal point in human travel arrived two centuries ago.
Huge changes unfolded between 1790 and 1830. At the start of
that period, most people traveled by carriage, riding along potholed
dirt roads at between four and six miles per hour. By all accounts it
was torture. If your route took you over the best roads, between
major cities such as New York and Boston, you could make the trip
in five or six days. You'd be hot or cold, beset by buzzing insects
attracted by the horses themselves, and it was not fun.

Two major improvements boosted long-distance transportation
to a new and celebrated average speed of between eight and nine
miles per hour. The first was the introduction of raised macadamized
roads with side trenches for drainage. This meant laying three courses
of stones, the largest on the bottom and the finest compacted at the

top. Riding on these "high" ways dramatically reduced lurching and bumping.[2]

The second speed booster was the stagecoach. By the 1830s, carriage companies used relays of horses that would be changed every forty miles or so along the route. With fresh horses attached at regular intervals, or *stages,* the New-York-to-Boston run was cut to one and a half days.

At around this time, steamboats increasingly plied waterways, starting with the *North River Steamboat,* soon popularly called the *Clermont,* which journeyed up the Hudson beginning in 1807, aided by the Erie Canal, which opened in 1825. Railroads (always called railways then) grew dramatically, too, and in the late 1830s they were routinely clocking in at between fifteen and twenty miles per hour. This was unprecedented, nonstop speed, and people paused from working in their fields to watch the wood-fired, smoke-belching, canopied carriages pass by, their occupants inhaling facefuls of soot and embers. By 1840, three thousand miles of track had been laid, mostly in the northeast, and that Boston trip now took a single day.

Children in the 1790s grew up to be astonished at the rapid change in travel speed they'd witnessed by the time they were grandparents in their fifties. It was a whole new world. There was a downside, however. As people increasingly voyaged by train and boat, roads were neglected and took on a rutted dilapidation by the mid-nineteenth century. They became suitable for local transportation only—the way you'd get into town from your farm or visit relatives a few towns over. This turnaround didn't reverse until the infatuation with the automobile took hold two generations later.

Cars were originally hailed as environmental saviors because they held the promise of eradicating the stench of horses, the thick swarms of flies and disease their feces attracted, and the unrelenting din of horseshoes on urban cobblestones. In today's Los Angeles and Beijing, few probably regard cars as "green" — but they do carry our story to the present, when we routinely hurl ourselves at seventy miles per hour along the freeway. Our very fastest body speeds? On the ground it's 180 miles per hour. That's the rate of European, Japanese, and Chinese bullet trains. It's also the takeoff speed of heavy jumbo jets just before they're airborne. It's the fastest most of us have ever moved on the ground.

In the air (a method of travel that marked a third major milestone), the speed depends on the jet. The normal, most efficient cruising speed of the good old Boeing 747 is 655 miles per hour. The newer giant double-deck Airbus A380 is a tad slower at 647 miles per hour, as is the Boeing 787.[3]

When we're not relying on technology to propel our bodies, our fastest movements are involuntary. One of these is legendary. Yet the *sneeze* usually begins in slow motion. The first phase of the sneezing reflex is a nasal tingling that follows stimulation by a chemical or physical irritant. Or sometimes by a strange brightlight response called the photic sneeze reflex, as when people emerge from a movie matinee into brilliant sunshine. Whatever the basis, the initial odd tingling grows until it reaches a level that triggers the far more animated second phase.

It's this so-called efferent phase that consists of eye closing, a sudden, uncontrollable deep inhalation, and then blowing out air while closing the throat and increasing air pressure in the chest. The reflexive sudden opening of the throat releases a supernova-type air rush through the mouth and nose, explosively expelling any irritants.

A sneeze can release forty thousand particles at high speed. What speed is it, exactly? You'll find all sorts of disparate velocity figures on the Web. Some claim that this is the only human-body movement that breaks the sound barrier. The truth, while still impressive, doesn't come close to such a 768-mile-an-hour achievement. The TV show *MythBusters* actually measured sneezes; their subjects' fastest was thirty-nine miles per hour. In a medical setting and using trustworthy equipment, the fastest recorded sneeze was clocked at 102 miles per hour. For some reason, the Guinness World Records lists the greatest sneeze a bit slower than this, at 71.5 miles per hour, or 115 kilometers per hour. Definitely fast enough to count as the highest-velocity body motion.[4]

A long-standing puzzle is why sneezers involuntarily close their eyes during the event. The best guess is that we are protecting our eyes from the ultrafast spray of germs and particulate matter. Another possible reason is that a sneeze is a unique reflex that involves nearly the entire body. Many muscles contract—including those in the nose, throat, abdomen, diaphragm, and back. Even the sphincters contract. This is why people with weak bladders may release a bit of urine during a sneeze. So the closing of the eyes is just part of a much larger, unique display of physiological violence.

It all originates in a primitive part of the brain called the medulla oblongata, in the brain stem, which is present in countless other animals who sneeze pretty much the same way we do.

So we just can't escape this hurry-up universe. We can't even avoid it by staying in bed.

We take it with us, inside our skulls and under our skin.

CHAPTER 12: *Brooks and Breakers*

Earth's Greatest Assets Are Liquid

But ol' man river,
He jes' keeps rollin' along.
—OSCAR HAMMERSTEIN II, "OL' MAN RIVER" (1927)

The headline was grim.

FIFTY-FOUR MIGRANTS DIE OF THIRST IN MEDITERRANEAN BOAT DRAMA.

Datelined Geneva, July 11, 2012, it recounted a horrific ordeal. Nearly five dozen migrants from Africa trying to reach Italy died of thirst when their inflatable boat ruptured in the Mediterranean, according to testimony from the sole survivor, Abbes Settou. The UN refugee agency UNHCR said that Settou, who drank seawater to survive, was spotted clinging to the remains of the stricken boat off the Tunisian coast by fishermen who alerted the coast guard. The man said there was no fresh water on board and people started to perish within days, including three members of his family.

It's the cruelest irony to die of thirst while immersed in water.

It also highlights water's critical importance. Of all the moving entities that surround and permeate our lives, the most vital are water and air—curiously, the only essentials that are transparent.

Our bodies are two-thirds water. Our brains are mostly made of it. No wonder these same brains enjoy watching it move as we dreamily stare at rivers and marvel at waterfalls. We bathe in water

168

and jump into it at the slightest provocation; it's the centerpiece around which vacations revolve. And, as with everything on this yin-yang planet, it sometimes turns on us, as my niece and Abbes Settou, sadly, learned.

Walls of water have always been terrifying. Yet aquatic fact and fiction competitively marched side by side for countless centuries. It took until passable science knowledge arrived in the nineteenth century before Noah's flood went from literal truth to mere parable. This happened only when it became obvious that if every ounce of water vapor in the atmosphere precipitated as rain, *it would raise the sea level by only a single inch*. No need for an ark. Noah's forty days of rain notwithstanding, floods, then as now, can never be more than regional events.

Still, the connection between water and humans may be even deeper than we suspect. Though generally ignored by anthropologists, the theory that Homo sapiens may be an *aquatic ape* linked genetically with lakes or the sea may explain such puzzles as our relative hairlessness, the size of our noses, and why we, unlike other primates, gasp when startled.[1]

In any event, Earth's existence as a water planet, where 70 percent of the surface is liquid to an average depth of twelve thousand feet, is unique in the solar system. But it's also logical, because H_2O is the most common compound in the cosmos.

This, too, makes perfect sense. The universe's most abundant elements are hydrogen, helium, and oxygen. Helium doesn't combine with anything, so cross that off the "most important" list. And even though oxygen is a thousand times less prevalent than hydrogen, it's always eager to join the party — any party. Small wonder that the H-and-O courtship and perennial "exchange of rings" is repeated in every corner of space and time.

Telescopes show water virtually everywhere. Steam envelops most stars. Comets are balls of dirty ice that turn into million-mile vapor streams, the stunning tail. Saturn's rings, among nature's grandest sights, are made up of countless chunks of ordinary ice.

The thing is, water stays frozen between minus 460 degrees Fahrenheit and plus thirty-two degrees Fahrenheit—and the temperature of most of the universe lies within this range. Water's gaseous state, steam, is maintained within an even larger temperature range, from 212 degrees to 2,700 degrees, when its molecules start to break apart. Thus ice and steam carry water's banner throughout the cosmos. The liquid form prevails in just a very narrow span, from thirty-two to 212 degrees.[2]

And even the right amount of warmth is not enough for water to maintain its liquid state. Though we mostly live in temperatures between thirty-two and 212 degrees, given that Earth's mean temperature is fifty-nine degrees, we still wouldn't see liquid water unless we were under pressure. In parts of the cosmos—such as Mars, during that planet's summer—enough heat exists to turn water into a liquid, but there is virtually no pressure, so H_2O remains exclusively vapor and ice on the red planet.

The weight of Earth's atmosphere provides the needed pressure. Reduce it and water readily boils off into a gas. All you need to do is drive up a hill to prove this for yourself. The boiling point plunges by roughly one degree for each five hundred feet you climb. So the hottest coffee in Denver is ten degrees cooler than the hottest coffee in Boston. Atop Mount Everest, water boils into vapor so readily that its maximum liquid temperature is about 160 degrees, unless one uses a pressure device. If astronauts hauled a bucket of water from their spacecraft to fill a decorative birdbath on the moon's desolate surface, it would boil furiously while simultaneously freezing solid.[3] Result: a grotesque modern art sculpture.

So H_2O is common, but its liquid form is rare. Yet this is what makes up most of our planet's surface as well as the human eyes that observe the whole pageant. We've got miracle stuff all around us.

A generous 326 million cubic miles of water cover the globe. Of this, 97.2 percent is ocean. Some 0.65 percent is in the form of freshwater lakes, streams, underground aquifers, and vapor and mist in the atmosphere. About 2 percent is locked in the form of ice. And all this liquid is poised to move the moment it finds a way to wriggle closer to the planet's center. It starts out by falling from the sky, making the water-flow scenario inevitable.

A truly immense amount of water is continually cycled through the air as vapor turns into plunging droplets. In a given year, 91,000 cubic miles of water fall as rain. If it all could collect on the surface, it would form a global layer nearly four feet deep. This is our world's annual rainfall, and it has to go somewhere.

Rain runoff begins as broad sheets that find narrow crevices or urban sewer drains in which to flow. From this point on, the water either seeps into underground pools or follows a channel whose width can vary from that of a narrow brook to that of the Amazon.

Streams and rivers can lope along at just half a mile an hour or race at twenty-five miles per hour, the fastest measured anywhere. Usually rivers flow at about walking speed; their average is three miles per hour. Even the Nile, during its famous yearly inundation, only rushes north at five miles per hour.

Rivers overwhelmingly do their damage in flood periods, when they're apt to overflow. Water is eight hundred times denser than air, so it decisively pushes anything it hits. It can move two-hundred-pound objects when it's just a foot deep. Still, much of a stream's erosion comes from the embedded particles—which in fast-moving scenarios can be entire boulders—that scrape away the sides.

After heavy rains the reliable runoff sequence starts with narrow rivulets that flow in channels and scour out steep-banked V-shaped streambeds. Over time, the sides erode and the channel becomes a wide, flat-bottom waterway. During nonflood periods the stream flows in the middle of this newly created valley.

Einstein seems to have been the first to point out that rivers tend to obey pi, the number 3.14159 and so on, with its never-terminating digits. Meaning that a river's straight-line distance from source to sea, divided into its actual meandering mileage over the ground, is pi. Rivers tend to build themselves a loopy path because the slightest curve will lead to faster currents on the outer side. This creates extra erosion and a sharper bend, which in turn produces a further increase of flow speed, accelerated erosion, and an even sharper twist to the river. (The removed sediment is typically deposited at the very next *inner* bend, building up the subsequent curve.) But a natural process limits water's desire for circuitousness: too much of a curve makes the river double back on itself, effectively short-circuiting the process by creating oxbow lakes. We're left with half circles and an overall value of 3.14. You can see this beautifully from the air or on a map.

The ratio of curved to straight sections varies from river to river. A ratio closest to pi is seen most commonly in rivers flowing across gently sloping terrain. In a river cascading down steep topography, the waters are too fast for the pi effect to work.

We'd expect to see this on other worlds as well, except no other place has rivers. Mars had flowing water millions of years ago, but no one's yet sure whether it stuck around for a long time or appeared only in sudden, temporary, white-water events. The fluvial channels along the Martian surface are ghosts from a long-ago era, and some appear curvy indeed. The only other liquid water within four light-years of the Mississippi is not the flowing variety but huge subsurface reservoirs. Jupiter's moon Europa and the Saturnian moon Enceladus boast warm, alluring saltwater oceans made possible by the weight and pressure of a mile of floating ice above them.

Sediments eroded and carried away by rivers add up to a staggering mass. Some consist of dissolved solids, such as salts, but the bulk is its *suspended load,* which is what makes a river turbid. It also transports a so-called *bed load*—material that moves along the channel floor by sliding and rolling.

The Mississippi alone carries 750 million tons of material to the sea each year. Two-thirds of this is in suspension (no surprise, given that river's famous chocolate color), two hundred million tons are in solution, and fifty million tons are in its bed load.

The flow rate is all-important. Water's kinetic energy (impact force) increases by the square of its speed. So when water doubles its speed, its talent for creating damage rises by a factor of four. During a flood, when river speed can easily increase threefold—say, from two to six miles per hour—its ability to scour its banks is multiplied ninefold. Very powerful. That's why such overwhelming damage and reshaping happen during floods.

Groundwater is usually on the move, too, but only by a few feet a day as it creeps through porous rocks or inches through cracks. Experts estimate that the hidden water within two thousand feet of the surface equals twenty times the volume of all the world's rivers and lakes combined. When it comes to fresh water, we only glimpse the iceberg's tip.

It is water, too, that most endangers us, even if only 5 percent of the 2,420,000 people who die annually in the United States succumb to *any kind* of a motion-based mishap.[4]

Most of these unintentional fatalities stem from car accidents or falls and such. Nature, by and large, can plead not guilty. All combined *nature-induced* fatalities account for just one death in a thousand. Nonetheless the pure drama of storm violence and the fact that *some* of us are indeed carried off by winds and earthquakes ensure headlines disproportionate to the actual peril. In a typical recent year it was water—floods—that killed one hundred Americans, while lightning killed sixty-five and tornadoes and other windstorms seventy-five. By contrast, car accidents killed thirty-six thousand people.

* * *

In the matter of aquatic motion, classical writers saved their reverence for the oceans. The ancients spoke of the seven seas, a term that gained household familiarity thanks to its use by Rudyard Kipling. Naturally, because all seas are interconnected, there's really only one global ocean, though salinities, currents, and other attributes vary from place to place.

Beyond the science is the sea's magic. Its size seems designed to induce humility. Watching it as it tirelessly moved and roared, Neanderthals stared at its swells, and so will the final humans, and even then its heartbeat won't falter. We cannot see air move or the galaxy rotate or the sun pulsate, but here at the shore Aristotle's "eternal motion" seems self-evident. The sea requires no intellectualizing.

What paradoxical sadism that this water should have slaughtered so many via the weapon of thirst. Desperation-induced seawater drinking first produces severe diarrhea, then confusion, brain damage, and finally death by renal failure. Drinking water with more than a 1 percent salt content rapidly increases blood sodium and pressure levels. The body reacts not to the water but to the salt, and the kidneys can remove it only by utilizing fresh water. Seawater is not even close to potable, being 3.5 percent salt by weight. (Typical city tap water has less than one hundred parts per million of sodium. The legal allowable maximum salt content is one thousand parts per million, or 0.001 percent.)

In bodies of water where evaporation is high but replenishment by rivers is low (e.g., the Persian Gulf and the Red Sea), the salinity can easily reach 4.2 percent. Conversely, the Baltic, into which a large amount of fresh water discharges, is just 2 percent salt. Ocean salinity is thus another action-based consequence of river activity. But enough beating around the bush: let's turn to the "big three" movers of the seas.

Waves. Tides. Currents. Each is epic. Each delivers untold tons of force.

* * *

Of all places to explore the tides, none competes with the Bay of Fundy in the Canadian Maritimes. I came here to see it firsthand, to stand atop a riverbank in Nova Scotia outside the town of Truro. The muddy bed of the Salmon River lay sixty feet below me. In its middle, an unimpressive stream a foot deep flowed leftward, toward a bend in the distance, and then presumably to the sea, unseen from this spot. An enterprising Canadian had built a restaurant on this promontory. Picture windows looked down toward the sandy abyss below. Nothing much, yet this is one of the most amazing places in the world.

Here is the location of a legendary *tidal bore*. At least, it's legendary among ocean lovers and people in the Canadian Maritime Provinces. Bores exist in just a few places on our planet, which explains why most folks have never even heard the sentence "It's a bore!" spoken with any kind of excitement.

The world-famous Bay of Fundy has a narrowing shape and, unseen beneath the surface, a precisely sloping seabed that together channel and amplify incoming tides. The Atlantic waters enter the fifty-mile-wide bay, and the constriction forces them to rise as they travel the 150-mile length. In the Minas Basin, near Wolfville in Nova Scotia, and here in Truro and a few other nearby places, truly bizarre consequences follow.

It all happens because, although the average coastal tidal range is five feet, here the sea frequently rises and falls sixty feet. Six stories. A *vertical* six stories. At high tide one sees floating ships, well behaved, tied to piers. A mere six hours later the boats are way below, sixty feet down in the mud, and the pier's odd full height, equal to a large apartment building, stands awkwardly revealed. As in a tidal wave's early, sickly, perilous stage, during which the unwary are lured into its grip, the ocean has retreated far off in the distance, separated from onlookers at the coastal road by a half mile of kelp and pools and happy gulls.

The world's tides fluctuate more dramatically when the moon and sun align, either together, at new moon, or on opposite sides of the heavens, at full moon. Twice a month, then, coastal communities experience these *spring tides,* whose name is confusingly awful, since they have nothing to do with spring or any other season. The term's origin is unknown. Perhaps people thought they were like a spring, a fountain of water. At these times, the high waters come nearly to the boardwalk, and the low tide exposes normally hidden stretches of muddy sand. This is when clammers, checking their tide tables, grab their buckets and shovels and head out. It's when the ocean typically rises and falls an additional foot or two beyond the average tidal range.

But here alone this is not the case. The strange, complex, watery wonders of the Bay of Fundy virtually ignore the monthly moon-and-sun alignments. Here the seas scarcely change during the time of spring tides. Instead, Fundy's tides increase when the *moon comes closest to Earth* — its monthly perigee. This changing-lunar-distance effect is small everywhere else. Here it matters mightily. That's why it's wise to check the lunar perigee tables before arranging any trip to Fundy if you want to witness the best tidal spectacles.

As I sat there I was a mile inland, with no sign of the ocean at all. Then, as if choreographed, people started filing out of the restaurant and standing on the raised embankment. All heads were turned to the left, toward that river bend a half mile away. People looked at their watches, their smartphones. There were hushed, expectant conversations.

And suddenly, there it was. The tidal bore. Rounding the bend like a living creature, and extending from bank to bank, a two-foot-high wall of water materialized and marched toward our position. When it arrived below us, roaring as its wave collided head-on with the river's flow in the opposite direction, its momentum carried their combined waters rightward. The ocean easily won the aquatic tug-of-war. The bore continued rightward until it was out of sight.

The show was not over. During the next hour, the river channel kept filling, higher and higher. It was the sea, marching farther inland, exploiting the low riverbed for its own advancement. By the time I left, the water was perhaps thirty feet deep, exhibiting a rapid march in the opposite direction from the way it was flowing when I'd first arrived.

Every coastal community has its own peculiar tidal sagas — albeit not as exceptional as they are here — for tides are often intricate and not always fully understood. Their origin is mainly lunar, though the sun exerts its own tides, a bit less than half the strength of the moon's.

Most people completely misunderstand what's afoot. The moon does *not* yank directly on ocean water. If it did, there might be something to the New Age belief that the moon's pull affects human lives — after all, our bodies are 65 percent water. Instead, the real story involves the moon's gravity in a very specific way. Because our cratered neighbor is so nearby, and because tidal forces vary with the cube (not the square) of distance between the earth and the moon, the moon exerts a greater "pull" on the side of the earth that's facing it than it does on the far side. This difference is not what causes the tidal effect. It *is* the tidal effect.

A tidal effect is not gravity, but the *difference* in gravity between two locations.

This is the critical point. For when the moon passes overhead, there is no effective difference in its distance to your head versus its distance to your feet, just five or six feet farther from it. The difference is basically zero. No *difference* means no tidal effect. Your bodily fluids remain in their customary locations.

But Earth's eight-thousand-mile diameter is another story. That's nearly 4 percent of the distance between the moon and the earth.

So the difference between the lunar force on Earth's moon-facing hemisphere versus its force on the opposite hemisphere sets up a bit of torque that results in a three-foot bulge of ocean water.[5] The moon mostly calls the shots when it comes to creating tides only because it's so nearby. In truth, the sun exerts a far greater *gravitational* pull on us—177 times more than the moon does. After all, it's twenty-seven million times more massive! But because the sun lies so far away, there's just not much difference between its strength on the opposite sides of our planet. And—I can't emphasize this enough—it's the difference that matters, not the overall gravity.

But tides are quirky. In Tahiti there are no moon-caused tides at all. French Polynesia only experiences a single daily *solar* tide of a paltry one-foot height. As the tidal bulge of water travels around the various seas, there's a rocking, an oscillation, and Tahiti happens to lie at a swivel point. It's like carrying a shallow pan of water. A back-and-forth sloshing quickly arises. But in the middle of the pan the water scarcely moves. Tahiti sits at that fulcrum spot in the Pacific. In other places, tides arrive at illogical times, thanks to the shape of the harbor or bay.[6]

Currents are the second mover of the seas. These are powerful rivers of seawater that have enormous influence. Ocean waters move continuously. Whenever we've swum or sailed in the ocean we've probably felt horizontal currents. Some come and go and shift with the wind or affect only a small beach area. But other currents can run through much of an entire hemisphere as a response to tropical heat and the prevailing wind.

Currents can flow anywhere from 0.5 to 5.6 miles per hour—generally the same as a river's speed. The Gulf Stream, which carries warm water from the Caribbean up the eastern coast of the United States and then to Europe, is one of the very fastest. Much more laid-back is the California Current, which brings chilly

Alaska water down past Oregon to San Francisco, making its beaches fit for seals and nobody else. Also slow is the famous cold Humboldt Current, which moves up western South America from Antarctica, letting penguins lounge closer to the equator than anyone would think possible.

About 40 percent of global heat transfer is carried by ocean-surface currents, which are generally less than a thousand feet deep. They are created and steered mostly by the prevailing wind.

Our final mover of the seas is waves—the most visually obvious of them all. Here, too, nearly all the energy comes from wind. Waves in the open sea are usually between five and fifteen feet high and run at forty-five miles per hour. It's important to remember that although a wave appears to be in motion, each individual drop of water does not move, except in a tiny circular path a few inches wide. After a wave passes, each drop of water is pretty much back in its original position. We see this clearly when watching floating debris.

Out at sea, waves are typically four hundred feet apart (the wavelength) and pass a given location every few seconds. In an individual series of waves, the interval between one wave and the next—sometimes as long as nine seconds but almost never more than that—never changes; the waves chug across the vast ocean in lockstep day after day.

All those days of lockstep monotony end when a wave reaches shallow water. As soon as its trough is half a wavelength's distance from the bottom, friction starts acting on the wave's base and increasingly slows it down. Meanwhile, momentum still carries its top forward at the previous rate. The result is that the wave's top rises while also leaning farther and farther ahead. When the steepness ratio reaches 1:7 (i.e., the wave's height is one-seventh of its length), it cannot support itself, and it "breaks."

It's stating the obvious that crashing waves exert enormous power. Their remorseless cycle of punishment lies almost beyond

Waves moving at forty-five miles per hour arrive from the open sea every five to eight seconds and "break" as soon as their height-to-length ratio reaches 1:7.

human appreciation. A single sea wave weighs thousands of tons. During storms, high waves can make the ground tremble with each impact, delivering a ton of force to each square foot of whatever substance—preferably inexpensive—receives its brunt.

Needless to say, the wave phenomenon reaches its terrifying extreme in a *tsunami*. Even in our own times, had it not been for the widely videotaped and heartbreaking events in the Indian Ocean in 2004 and in northeastern Japan in 2011, a tsunami would still be popularly misconstrued as a single looming tidal wave moving toward shore. Now few would make that mistake. Whereas the average normal ocean wave travels at forty-five miles per hour, a tsunami moves at around five hundred miles per hour, rivaling a jet aircraft. And yet, the Noah story aside, the ancient world, too, seemed largely unaware of the possibility of the ocean utterly changing its behavior and taking countless lives.

It is true that, in prehistory, a fierce tsunami devastated the Norwegian Sea around 6000 BCE, but no records existed to warn Middle Eastern, Persian, and Mediterranean civilizations of tsunamis' devastating power. And nearly the entire island of Thera, now called Santorini, was destroyed by an explosive eruption around 1650 BCE that created a tsunami so strong it wiped out the advanced Minoan civilization on nearby Crete. Yet a millennium later, when parts of the Bible were penned and the first thinkers of classical Greece were observing nature, this, too, had been essentially forgotten. A thousand years is, after all, a long time.

This lack of awareness changed in the summer of 426 BCE, when a modest tsunami startled the sailors in the ships armed for the Peloponnesian War. In his written history of that conflict, the Greek historian Thucydides openly mused about what could possibly cause the ocean to behave so strangely and correctly concluded that it must have been an undersea earthquake. He thus became the first to link the movement of solid earth with that of the liquid seas.

A half century later, in 373 BCE, a tsunami permanently submerged the town of Helike in Greece, obliterating its population and perhaps inspiring Plato, who was in his midfifties at the time, to speculate about a lost civilization that he called Atlantis. And yet this, too, was a largely localized event.

The same was not true 738 years later.

On July 21, 365 CE, an enormous undersea quake, the likes of which occurs only every few thousand years, struck the eastern Mediterranean between Crete and Egypt. Although everyone in the region felt the ground shake violently, it passed without widespread destruction—except on Crete, which received no warning that an astonishing hundred-foot-high wall of water was radiating outward. When we recall the eighty-foot tidal wave of the 2004 tsunami that killed a quarter million people in places such as Banda Aceh, Indonesia, or the seventy-seven-foot-high tsunami that wiped out Japan's Fukushima Daiichi nuclear power plant in 2011 (thanks to the back-up diesel generators having been sited, bewilderingly, on the ground floor), we can appreciate how horrible a hundred-foot—ten-story—wall of water must be.

We have an actual eyewitness account of the 365 CE tsunami from a survivor. And not just any survivor but the Roman historian Ammianus Marcellinus, known for his accurate, unembellished accounts of the everyday life of his time. It was he who watched in astonishment as an event that was anything but everyday unfolded and recounted it almost matter-of-factly in Book 26 of his epic, *Res Gestae:*

> Slightly after daybreak... the solidity of the whole earth was made to shake and shudder, and the sea was driven away... and it disappeared, so that the abyss of the depths was uncovered and many-shaped varieties of sea creatures were seen stuck in the slime.... Many ships, then, were stranded as if on dry land, and people wandered at will... to collect fish and the like in their hands; then the roaring sea... rises back in turn, and through the teeming shoals dashed itself violently on islands and extensive tracts of the mainland, and flattened innumerable buildings in towns.... For the mass of waters returning when least expected killed many thousands by drowning.... [H]uge ships, thrust out by the mad blasts, perched on the roofs of houses... others were hurled nearly two miles from the shore.

Another historian, Thucydides, said that "without an earthquake it does not appear to me that such a thing could happen."

He was not entirely correct. Any large mass, such as a meteorite, hitting the sea can displace enough water to do the job. The largest wave ever recorded, an astonishing 1,720 feet tall, or about 50 percent higher than the Empire State Building, raged through Alaska's Lituya Bay on July 9, 1958. This was the tallest tsunami in history. It did begin with an earthquake, but not a particularly big one; yet the tremor knocked loose a mass of rock that plunged three thousand feet into the Gilbert Inlet, displacing enough water to create the monster wave. The sheer scale of such motion defies visualization. After all, each cubic mile of the ocean weighs five billion tons.

The 365 CE tsunami went on to annihilate much of Alexandria, Egypt; Crete; and coastal Libya, and marched up the Nile River delta, hurling ships two miles inland. The quake permanently raised the coast of Crete by thirty feet — to this day the record elevation gain resulting from a single sudden event. This sunrise tsunami was so devastating that its anniversary was commemorated each year in Alexandria as "the day of horror" — *for the next two centuries!* It was quietly forgotten only at the end of the sixth century.[7]

Observing waves is everyone's idle pastime. Waves have many cool attributes, as people such as Otis Redding have perenially noticed while they're "sitting on the dock of the bay, watching the tide roll away." One favorite attribute involves *diffraction.* We experience this principle when we turn on the car radio and the FM stations fade in and out as we pass close to hills or buildings. But the AM signals are much steadier and don't vanish so readily. This is due to the bending of electromagnetic radiation around obstacles — diffraction. Longer wavelengths diffract more readily. AM stations broadcast waves hundreds of times longer than those on the FM band, so they bend around obstacles much more easily and are therefore not as readily blocked by obstructions. In other words, we don't get into radio *shadows* as easily when we listen to AM's widely spaced waves.

Now back to the ocean, whose waves are also pretty long—hundreds of feet—so they're not easily stopped by a small obstacle. Notice how waves encountering a little rocky lighthouse island soon fill in again behind the island and continue on their way. If the sea waves were closer together, there'd be more of a "shadow" zone behind small islands in which the ocean is permanently calm. By noticing this phenomenon behind jetties, docks, and other obstructions of various sizes you can see the diffraction effect in action.

Major mysteries of maritime motion remain. For example, south of New York City, much of the coastline as far down as Florida is peppered with barrier islands—low offshore sand ridges that run parallel to the coast. Waves crash onto them, and thus they shield the tranquil lagoons behind them, where boaters enjoy miles of protected sailing without having to take to the rough open sea.

The question is: Why should barrier islands endure? All shorelines suffer major modifications from the relentless pounding of the sea and storms. Logic tells us that these barrier islands should not last long. Their continued existence is a mystery, although this doesn't stop oceanographers from guessing.

Do breakers heap up sand scoured from the bottom of the sea and continually deposit it on the shore, replenishing the islands? Their sands are much higher than the high-tide mark, so any sand must be deposited during extreme storms. But such storms might just as well wash away these long, narrow, fragile islands, so we're back to square one. Or are the islands perhaps remnants of giant sand dunes, maybe deposited during the last glaciation period? If so, then perhaps it's the calm sea between them and the mainland that requires explanation—does it conceal a lowland running parallel to the dune ridges that got submerged when the sea level rose?

This lone example—and one could easily find hundreds—illustrates that even some simple aspects of the powerful moving sea and its relation to the long-suffering coastline remain uncertain. Our current science is not always up to the task of fully appreciating the ceaseless aquatic pageants.

Perhaps, as so many before us did, it's sometimes better to simply sit "on the dock of the bay," wasting time by watching the waves.

CHAPTER 13: *Invisible Companions*

The Odd Entities Zooming Through Our Bodies

Yesterday, upon the stair,
I met a man who wasn't there...
— HUGHES MEARNS, "ANTIGONISH" (1899)

Before the twentieth century, most people believed in ghosts or spirits. Yet no one in all of history suspected that *tiny* invisible entities zoom right through our bodies 24-7. Expressing such a belief would have gotten you thrown into a medieval insane asylum, where they probably didn't even accept MediSerf.

This invisible stuff is part of our everyday lives. It's not entirely harmless. Yet there's nothing we can do to get rid of it unless we ask a real-estate agent to find us a nice two-bedroom deep in a mine.

This story begins in 1800. That's when William Herschel discovered a form of light nobody can see. Invisible light? If anything ever came from left field, it was this. It fit nowhere in mankind's evolving models of the cosmos. The discovery would have surely been doubted and ridiculed except that Herschel was then the world's most respected scientist, famous for having found the first-ever new planet, Uranus, nineteen years earlier. (Nobody had seen *that* one coming, either.)

Since light can be regarded as a stream of particles, we can truly say that countless unseen bullets continually zoom around us. This first-known invisible form of light doesn't quite go unnoticed, though. Our skin detects Herschel's "calorific rays" (eventually

186

called infrared radiation) as the sensation of heat. Nearly half the sun's emissions are infrared. So when we look around us, an equal mix of visible and invisible particles are bouncing off the rocks and rabbits.

You may imagine that heat moves slowly. It takes a while to warm up a frying pan. But infrared rays, which create heat on our skin by making its molecules move faster, are light-speed swift. You experience this when gathered around a campfire on a chilly night. If a big person steps in front of you, you instantly feel the effect because that person blocks the invisible infrared rays from hitting you. He's creating infrared shadows.

A year after Herschel's discovery, in 1801, the sad-sack German Johann Ritter discovered ultraviolet light but failed to publicize it sufficiently. He had a tendency to ramble on about extraneous matters, such as his belief in ghosts, so he ended up ignored and impoverished. He wasn't credited with his discovery until after his death—the glory, appropriately, awarded to his disembodied spirit.

Things took an even more disquieting turn near the end of that century. On November 8, 1895, another German—Wilhelm Röntgen—discovered X-rays. As we all know, these waves or particles do not stop when they reach the skin. They can fully penetrate our bodies, although many are absorbed by dense material such as bones and teeth. When, two weeks after his discovery, Röntgen took the very first X-ray pictures, showing the hand of his wife, Anna Bertha, she stared with horror at the image of her skeleton and exclaimed, "I am seeing my death!" (Given the then-unknown deadly potential of shortwave radiation such as X-rays, which ultimately took the life of Marie Curie—the first person to win two separate Nobel Prizes—and many thousands more in places such as Chernobyl and Hiroshima, Anna Bertha's comment may seem eerily prescient.)

In 1896, the Dutch physicist Hendrik Lorentz posited the

existence of a totally different invisible speedster: the first-ever subatomic particle, the *electron,* even tinier than the theoretical atoms suggested by Democritus 2,300 years earlier. Lorentz had plunged deeper than any other physicist before him and brilliantly figured out the origin of all light! He said that light comes into existence solely because of the motions of a tiny, negatively charged object. When the electron was duly discovered soon thereafter, Lorentz's prescience earned him the 1902 Nobel Prize in Physics.

This was a productive time for finding invisible entities. The pace didn't let up. Also in 1896, French physicist Henri Becquerel got swept up in the global excitement of Röntgen's X-ray discovery of the previous year. Becquerel's interest was in materials that glow, so he thought phosphorescent substances such as uranium salts might emit X-rays after basking in sunlight. By May of that year, however, he correctly realized that the uranium emitted some new and unknown form of "radiation," as it was starting to be called. Seven years later, in 1903, Becquerel won the Nobel Prize in Physics, sharing it with Pierre and Marie Curie, who had taken Becquerel's ideas and run with them.

The newly married couple was fascinated by substances that emitted what Marie called uranium rays. After years of watching these strange rocks produce smatterings of light on photographic film, the Curies realized that the most intense radiation flew out from two brand-new elements. She named the first polonium, after her native land, Poland, and the second radium, for the mere act of radiating. This latter element was her baby, her darling; she called it "my beautiful radium," for she possessed no inkling that it would someday kill her and many others with its sizzling emissions. It was three thousand times more radioactive than uranium.

So now, quite suddenly, nineteenth-century scientists had revealed a motley crew of five invisible entities flying around or through us.

Ultraviolet photons can burn us at the beach and set the stage for skin cancer, but they are also beneficial, even vital; the body creates vitamin D when struck by them.

X-rays are scarcely present naturally here on Earth.

Infrared rays are commonplace but harmless.

So are electrons, streams of which were used for decades in the old-style TV picture tubes to conjure Mister Rogers and Lucy Ricardo.

But Becquerel's and the Curies' uranium- and radium-based "radiation" would prove far more dangerous, even if radium was initially believed to be a healthful substance, a tonic. (To this end, it was marketed as an elixir, mixed with sparkling spa waters and touted as a rejuvenating agent. Millions of bottles were sold and drunk. Later came radium watches with their glowing numbers and dials, painted in factories mostly by young women who suffered horrible early deaths before the peril was recognized.)[1]

The spooky quest for unseen phantoms soon got even spookier. In 1909 Theodor Wulf created an early equivalent of a Geiger counter — an instrument called an electroscope, which revealed whether atoms inside a sealed container were being broken apart. It showed higher levels of radiation at the top of the Eiffel Tower than at its base. Because this made no sense — the device was then farther from the ground's uranium and radium sources — his paper was ignored. But on August 7, 1912, Austrian physicist Victor Hess personally took improved versions of the electroscope up in a hydrogen balloon to 17,400 feet, and it revealed radiation levels twice as intense as those on the ground. He correctly attributed this to a radiation source arriving from outer space.

Hess soon eliminated the sun as the cause: he flew a balloon during a solar eclipse, when the moon blocked nearly all the sun's incoming energy. He also, perilously, conducted some flights at night. The conclusion was amazing, if disquieting. He announced, "A radiation of very great penetrating power enters our atmosphere from above." For this discovery—which still has ominous ongoing implications for pilots, in addition to posing a serious hazard to any future human colonies on other worlds—Hess won the 1936 Nobel Prize in Physics.[2]

By amazing coincidence, precisely one century to the day after Hess's balloon flight, on August 7, 2012, the newly landed Mars rover *Curiosity* began measuring this radiation on another planet for the first time.

Physicists initially believed these invisible outer-space invaders were some kind of wave, an electromagnetic phenomenon, which is why they were—and mostly still are—called *cosmic rays*. Each of those two words tingles with sci-fi creepiness and vaguely implies a bizarre peril from beyond the stars, thus awarding cosmic rays the scariest and perhaps coolest name of all the tiny, streaking, high-speed entities.

But they aren't rays at all. Meaning they're not a form of light. Their incoming paths are bent by our planet's magnetic field, and light never changes direction in response to magnetism. Cosmic rays simply couldn't be another electromagnetic phenomenon, as X-rays are. Before the start of World War II, everyone realized they must be electrically charged particles, like the ones that stream (as we finally recognize) from uranium and radium.

The truth, the denouement, is both powerful and anticlimactic. Cosmic rays are mostly protons. Ordinary, plain-vanilla protons, the nucleus of hydrogen, the positively charged particle found in every atom's heart. They're violently ejected in supernova explosions and wander the universe like homeless high-speed swashbucklers.

But why is 90 percent of this incoming substance protons? Why does it include just a negligible sprinkling of electrons (1 percent)? There are just as many electrons as protons in the universe. Why are electrons so underrepresented?

Befitting their spooky name, cosmic rays are thus puzzling even

190

today, thanks to their illogically proton-heavy composition and the fact that a small percentage of them scream into our atmosphere at bewilderingly high, near-speed-of-light velocities. Cosmic rays even include a bit of antimatter.

Protons weigh 1,836 times more than electrons, so they pack a wallop when they hit anything. Fortunately our atmosphere and our magnetic field block most of them. While you and I do get penetrated regularly, they're a medical problem mostly for astronauts, which is why the twenty-seven Apollo adventurers all saw spurious bright streaks cross their visual fields every minute, as protons ripped through their brains.

Moreover, as in a game of billiards, protons typically strike air atoms thirty-five miles up and knock loose a cascading shower of smaller stuff. One of these is the muon, which decays rapidly but not before it, too, penetrates our poor pincushion bodies.

At least two hundred muons per second zip through each of us. They weigh 208 times more than electrons, so they're not exactly harmless if they crash through and alter a gene in one of our chromosomes. You can avoid them only if you live underground, in a place like Zion, the city in the Matrix films. The mutations they induce keep plants and animals evolving and help explain why today's cats and cabbages look different from their analogues from a hundred million years ago.

It's now clear: we're continually hit and penetrated by many different unseen particles and waves. Some of it is harmful. If this worries you and you tell a therapist of your concerns, you will be advised to come more regularly. To which you might then suggest that future sessions be held in an underground parking garage.

Nowadays the blanket term *radiation* can mean any kind of invisible high-speed detritus, but usually the word pertains to just those that can produce genetic defects and cancer. They include short waves, such as X-rays and gamma rays, as well as solid particles, such as those cosmic-ray protons and the even more massive alpha particles.

The universe is filled with radiation. It's everywhere. It comes up from the ground and down from the sky. Most people are clueless about it. They don't even know what radiation is. They don't grasp its dangers. Yet we can easily calculate our personal yearly exposure.[3]

We measure exposure in millirems. Except for those who get periodic medical CAT scans, the average person gets 360 mrems a year, of which 82 percent comes from natural sources—even when we're far from any health food store. This radiation is responsible for some of the spontaneous tumors that have always plagued the human race.

Our atmosphere blocks some of it, but the higher up you live, the more you get.

Nonetheless, and curiously, Tibetans and Peruvians, who spend their lives at high altitudes and therefore receive much more radiation than folks in Newark, New Jersey, *don't* suffer increased rates of leukemia. And a major 2006 French study showed no increased cancer incidence in children who live near nuclear power plants. These, however, are minor radiation sources. They are dwarfed ten thousand times over by the "top three," which we will get to in a moment.

If you're concerned about all this high-speed detritus zipping through your favorite organs, here's how you can calculate your personal annual exposure.

First the biggies:

Award yourself 26 mrem just for living on the surface of the earth.

Add 5 mrem for each thousand-foot elevation of your home. If you live in Denver you have to add 25 mrem, since you're closer to those cosmic rays.

Is your home stone, brick, or concrete? Anything but wood frame? Add 7 mrem. These materials are slightly radioactive. Your real-estate agent probably never mentioned this.

Do you have a below-grade basement? If radon is present, add at least 250 mrem; this is a genuine biggie. The average homeowner gets most of her yearly radiation this way— from radon. It's the densest gas you're ever likely to encounter. It therefore likes to accumulate on your lowest floors. It is also the only gas that has only radioactive isotopes, so it's a health hazard, no questions asked.

Radon is created when uranium and thorium decay, and these elements lie beneath many houses. Here's something odd: the decay of radon gas itself produces new radioactive elements that are always solids. These stick to dust particles in the air and get inhaled into lungs. This is the single greatest cause of lung cancer after smoking. But some homes have none of it. An inexpensive test can let you know. In basements that do emit radon, it tends to accumulate, though it is easily remediable by the use of venting fans.

Add 40 mrem for the radiation you get from food and water. This is unavoidable.

Award yourself 50 mrem for the natural radiation emanating from within your own body—from potassium, for example, if you're fond of bananas. Yikes! Eat a single banana and you've received more radiation than your friends who live next to a nuclear power plant get in an entire year.

Add 1 mrem for radiation in the air left over from those atomic tests in the 1950s. This, too, is unavoidable. You knew those bastards were screwing around with everyone's children's future health. We are those children.

Now for the major radiation sources you *could* avoid:

Add 1 mrem for each thousand miles you travel by jet this year. A single round-trip from Washington to Los Angeles gives you 6 mrem. That's why the professional pilots and crew who get all this daily radiation have a 1 percent greater rate of cancer than the rest of us. Their rate is twenty-three cases per hundred people instead of twenty-two for the general population. They don't tell this to prospective pilots taking flying lessons.

Add 40 mrem for each medical X-ray you undergo.

Add an astonishing 1,000 to 5,000 mrems for a whole-body CAT scan. Some machines deliver as much as 10,000 mrems. Given the sixty-two million CAT scans performed annually in the United States, this has probably replaced radon as our greatest single radiation source. *One CAT scan can give you more radiation than that received by Hiroshima survivors a mile or two from ground zero* (about 3,000 mrems, on average). It's reliably estimated that about 2 percent of cancers in the United States are caused by CAT scan irradiation.

Add 1 mrem if you watch TV on an old-fashioned tube-type set.

Obviously, testing your basement for radon, installing a venting fan if necessary, and asking your doctor if an X-ray might work as well as that CAT scan are the easiest ways to *greatly* reduce your radiation exposure. Avoiding unnecessary commercial flights could cut down a little bit more. It presents a tangible reason not to visit the in-laws.

Finally, here are the minor sources of radiation that give you less than 1 mrem per year, presented in decreasing order of dosage. These are the things you can truly forget about:

Looking at a computer screen: 0.1 mrem.

Wearing an LCD watch: 0.06 mrem.

Living within fifty miles of a coal-fired power plant: 0.03 mrem. (That's because coal and soot are slightly radioactive.)

Having two smoke detectors in the house: 0.02 mrem.

Living within fifty miles of a nuclear power plant: 0.009 mrem.

Having luggage X-rayed once: 0.002 mrem.

Going through an airport's backscatter X-ray scanner: 0.01 mrem.

Can such superlow doses produce any harm at all? The Mayo Clinic, the Health Physics Society, and most epidemiologists believe there's a threshold below which radiation is as benign as popcorn. Disputing them, some scientists believe that very low doses in the 1 mrem range might create some small effect along the lines of one cancer death per forty million people. Even this minority group of enhanced worriers agrees, however, that the risk involved in having two smoke detectors or living near a nuclear plant is *so* low that it poses no risk whatsoever (barring an accident, of course).

If radiation concerns you, don't even *think* of moving to Mars. Martian colonists might receive enough radiation in two years to destroy 13 percent of their brains. Some say 40 percent.

Radiation and other subatomic particles and photons aren't the only things flying through us. There is one additional standout that dominates everything else. Indeed, the most abundant thing in the universe—besides light itself—is the *neutrino*.

Neutrinos are as omnipresent as roaches in Rio. Each second, two trillion neutrinos fly through each of our tongues. Don't taste a thing? That's because they rarely meddle with our bodies. Despite their unrelenting, torrential presence, they are utterly harmless.

The existence of the neutrino was predicted eighty years ago to explain odd atomic behavior involving the neutron, whose name it resembles. But don't confuse them. Neutrons live in the center of all atoms except the most common form of hydrogen. They're the heaviest stable particles and live forever. Although here's something weird: if a neutron leaves its atom, it's toast. It then decays in about eleven minutes.

A loose neutron is a loose cannon that goes *poof* and turns into a proton and an electron. This detritus arcs away oddly, like defective fireworks, which persuaded Wolfgang Pauli in 1930 that something else must be present, too, something that weighs almost nothing. The mystery item was soon named the neutrino, which means "little neutral one." A quarter century passed before the existence of the neutrino was finally confirmed. It was a science triumph and a relief, since neutrinos already figured theoretically in the fusion process, which makes the stars shine. The sun's core releases countless neutrinos that essentially zoom away at the speed of light and do not interact with anything else, at least not in a normal way. Rather, they just pass through everything.

By day, neutrinos from the sun penetrate your head and shoulders and whiz completely through your body and then proceed into and through our planet and out the other side. At night, an equal number invades your body from below and exits through your head, after having flashed through the entire earth in one-twentieth of a second, as if our planet were no more substantive than fog.

The chance that a neutrino will alter even one of your body's seven octillion atoms anytime this year is one in a million. You need a wall of lead a light-year thick to stop the average neutrino.[4]

Lately we've found that neutrinos are even stranger than we imagined. Credit for this goes to physicist Ray Davis, who first figured out how to count these ghost particles. The apparatus he used is a huge vat of one hundred thousand gallons of dry-cleaning solvent. Davis placed this a mile underground, in the abandoned Homestake gold mine in South Dakota, penetrable only by neutrinos. Even bats can't get there. He figured a million pounds of perchloroethylene has so many chlorine atoms that a neutrino will occasionally change one to a form of argon. He managed to detect the trace of six neutrinos after four months — that's how rare it is that a neutrino will bother anyone or anything, even an atom among countless trillions. That was forty years ago. That his brainchild actually worked earned Davis the 2002 Nobel Prize four years before his death.

Neutrinos, which come in three different varieties, change from one form to another as they fly through space. But the bottom line is that they are absolutely everywhere. Your thumbnail is penetrated by a trillion neutrinos every second. But because they do not normally influence ordinary baryonic matter, they do not cause gene mutations and, hence, cancer. They're like dust mites, our harmless sleeping buddies. But nobody develops allergies to neutrinos.

QUICK GUIDE TO FAST INVISIBLE OBJECTS STRIKING YOU RIGHT NOW

…and whether they're safe (S), harmful (H),
or pose just a slight risk (SR)

Infrared photons	S
Ultraviolet photons	SR
Electrons	S
Cosmic rays	SR
Neutrinos	S
Muons	SR
Dark matter	S
Alpha particles (e.g., radon)	H*

From a psychological standpoint, neutrinos pave the way for the most recently posited unseen entity: dark matter, our final zooming phantom. This is today's most mysterious substance.

It's been obvious since 1933 that there's six times more material in the universe than all the stars, nebulae, black holes, planets, cheeseburgers, and everything else we can think of combined. This unseen stuff makes the Milky Way spin strangely. It glues together the Local Group of galaxies. Its gravitational pull is powerful. Yet it's invisible. We call this material dark matter. Each of its particles must be massive.[5]

Because the universe is definitely filled with "presto chango" neutrinos that barely affect anything, dark matter may simply be a

---

*Half of all radon gas decays in 3.8 days into tiny solid radioactive entities that emit alpha particles, which are breathed in and pose a lung cancer risk. The quantity matters mightily, which is why all homes should be tested for radon.

heavier version of these. But because it does not emit or even reflect light, it must have nothing to do with electrons.

I had witnessed astronomers' obsessions with dark matter that night atop the Chilean mountain. I had watched astrophysicist Barry Madore scrunch his brow while studying a fresh photograph he had taken with the hundred-inch telescope. The galaxy's features defied logic.

"We still don't understand these dark matter halos around galaxies," Madore had muttered to me. "Look at these ragged edges," he said, tapping a finger on the photograph's outer section. "What's doing this? Why is this material flying off into intergalactic space? What's causing this motion?"

He finally shrugged.

"My educated guess is that it's dark matter."

It seems to pervade the universe. It may even lurk, unseen, in the rooms of our homes, in the very air around us. A study conducted in 2012 showed that it is not confined to galaxies but bleeds off into the seemingly empty space between them. It may be everywhere. Another 2012 study revealed them to thickly inhabit the region near the sun and its neighboring stars.

Dark matter is detectable not by its appearance—for it has none—but by what it does to everything near it. Its gravitational attraction glues each galaxy cluster together, keeping its members from wandering their separate ways.

Madore stared at the image again. "But maybe gravity itself acts strangely when it's far away from whatever it's tugging on. So maybe there is no dark matter at all. I mean, where's Occam's razor here?" he concluded, referring to the principle that the simplest explanation is usually the correct one.

But which is simplest? The idea that gravity behaves weirdly at long distances, a theory called MOND, which is embraced by only a minority of researchers? Or the concept of a bizarre new form of matter? Barry Madore had sighed loudly and asked, "Which do you choose, the improbable or the exotic?"

It's even possible that these particles *attract each other* to form invisible structures. Could there be a major parallel universe right here among us?

Most of our bodies and our planet and indeed every atom everywhere is utter emptiness. If the universe were compressed and all its spaces were removed, you could squeeze everything that exists into a ball smaller than a supergiant star, such as Orion's Betelgeuse. It's not totally far-fetched that these wide-open spaces in our bodies allow for cohabitation with creatures or objects of some other realm. Perhaps—since we're now letting ourselves speculate without a shred of supporting evidence—conscious entities whiz through our everyday lives like ghosts, as oblivious to us as we are to them.

Recent studies from 2012 and 2013 that looked at the "halo stars" in our galaxy—those high above the flat plane where the vast majority of suns dwell—show that nothing seems to be tugging them. Yet those locations are just where dark matter was thought to predominate. On the other hand, studies of the motion of galactic clusters—distant large objects—do indeed indicate dark matter's existence. In short, today's evidence is bewildering and contradictory. And unlike other zooming entities that continually penetrate our bodies as if we were Swiss cheese, dark matter is undefined: we still can't say what it is.

Maybe we're better off not knowing.

CHAPTER 14: *The Stop-Action Murderer*

And His Battles with the Ephemeral

Noise proves nothing. Often a hen who has merely laid an egg cackles as if she had laid an asteroid.
— MARK TWAIN, *FOLLOWING THE EQUATOR* (1897)

Breezes, crickets, lava, digestion, spinning moons, diving birds, blowing sand — we've mostly explored movement that unfolds in familiar time frames. Stuff we can see, things perceived even before the word *science* was coined. But starting with the ancient Greeks, long before the era of stop-action photography, observers grew intrigued by *ultrafast* entities. Like the beating of hummingbird wings 1,250 times a minute, these events were not mysterious so much as *fast to the point of invisible.*

People knew of them by the things they left behind. Perhaps there was a whine, as of mosquito wings, or a shadowy blur to mark the lower terminus of an intriguing action too swift to see.[1]

This unperceived universe was captivating from the get-go because it involved speed, prized by many cultures as a desirable attribute in people and animals. The very fastest actual animal in any of the constellations the Greeks identified — as opposed to mythological animals, such as Pegasus, the flying horse — was the big dog, Canis Major, whose hastening legs, according to legend, won his epic race against the fox, supposedly the world's swiftest animal. That victory was enough for Zeus to immortalize the dog in the heavens. (In a real foxes-versus-dogs derby, the result would be close to a photo finish; they've respectively been clocked at forty-two and forty-four miles per hour.)[2]

Blurry legs and wings and other whizzing things intrigued Renaissance scientists. Several desperately tried to study this invisible realm of speed. By rapidly waving fingers in front of their eyes, some created a crude "stop-action" strobe effect that could actually reveal a hummingbird's wings in flight, even though they flapped twenty times a second. If you try this in front of a spinning fan and vary your hand-waving rate until it works perfectly, you'll actually be able to freeze the blur and see individual blades clearly.[3]

More sophisticated techniques finally started uncovering nature's high-speed secrets in the late nineteenth century. To this day, one person alone remains famous for making such invisible motion decipherable. He was the bearded mastermind now known as Eadweard Muybridge.

Muybridge's 1878 sequence of photographs of a galloping horse, shown as a repeating loop, is the most famous "movie" of the nineteenth century. By "slowing down" previously too-fast-to-perceive action, it settled a longtime debate about how horses run, a nagging issue in the 1870s, when those ubiquitous animals dominated the urban as well as rural landscape.

Born Edward Muggeridge in England, this singular and not always likable character kept changing his name after emigrating to San Francisco in 1855, at the age of twenty-five. He started calling himself Muygridge, later changed his first name to Eadweard, and yet signed all his photographs "Helios." Moreover, when doing photo shoots south of the border he insisted his name was Eduardo Santiago. (His gravestone gives yet another name, Eadweard Maybridge, so even now it's hard to know *what* to call him.)

At the start of his career he was a bookseller and an agent for a British publishing company at a time when San Francisco had dozens of bookstores and almost as many photography studios. But his life changed in the summer of 1860, when he embarked on a journey back to England that started out via the southern overland route to New York. The trip ended violently in Texas.

Muybridge suffered severe head injuries in a splintering stagecoach crash that killed one man and badly injured everyone else on board. After spending three months in treatment in Arkansas, he awoke to remember nothing of his earlier life: his memory began anew at this point.

Muybridge continued treatment — for blurred vision — in New York over the course of the following year. He also had permanently impaired senses of taste and smell and exhibited erratic, emotional, and eccentric behavior. A few biographers have since claimed that the apparent damage to his frontal cortex actually freed him from inhibitions and paved the way for the motion-related photography breakthroughs, which ultimately made him famous.

He finally resumed his trip to England, where he underwent further treatment. There, Muybridge studied the newest photography techniques — and improved on them. He soon was granted two patents for photo-related inventions.

When he returned to San Francisco in 1867, he was no longer a publisher's agent and bookseller but rather a professional photographer with cutting-edge skills and an artistic talent none of his friends had previously seen. He quickly gained wide renown.

His big break came in 1872, when former California governor Leland Stanford, a wealthy racehorse owner, asked Muybridge to undertake a very specific photographic study. This involved an issue debated endlessly in racing circles: whether all four feet of a horse are ever off the ground at the same time while the animal is trotting or galloping.

No one could tell by simply watching. Artists had always painted galloping horses with either one foot on the ground or all four feet in the air simultaneously—often with the front legs extended forward and the hind legs extended to the rear. Stanford wanted to know once and for all. He offered Muybridge a handsome sum if he could settle the debate.

Muybridge did not fool around. He took a series of photos in 1878 by positioning numerous glass-plate cameras in a line along the edge of a track. The horse would trigger each camera in sequence by hitting a thread attached to the shutter as it passed by. Muybridge hung up white sheets behind the cameras to reflect the maximum amount of light during these stop-action brief exposures. Later he assembled the photos as silhouettes onto a spinning disk that fit easily onto a tabletop. The viewer would look at the spinning disk through a slit and see one photo at a time. It imparted the striking illusion of smooth motion. Muybridge invented this device, which he called a *zoopraxiscope*.

It not only clearly revealed the running horse's gait but also became all the rage. *Scientific American* did a story on it in which Muybridge was portrayed as a modern Isaac Newton. (History regards the device as so groundbreaking that in 2012, on its one hundredth anniversary, Google ran a continuous repeating Google Doodle of the "movie.")

Soon called *The Horse in Motion* (or, alternatively, *Sallie Gardner at a Gallop*), it is not only the world's first bit of cinema but also opened the floodgates for others' stop-action images, which uncovered the hidden, blurry universe of ultrafast physical events. The zoopraxiscope itself was the main inspiration for Thomas Edison's first commercial film-viewing system, the kinetoscope.

As for the horse-gait debates, Muybridge's sequence reveals not just a galloping horse with all four feet off the ground. It also shows that this fully airborne moment did *not* happen when the horse's legs were extended to the front and back, as depicted by eighteenth- and nineteenth-century illustrators. Rather, a horse is fully airborne only with all its legs bunched beneath its body, during the moment when it is transitioning from "pulling" with the front legs to "pushing" with the back ones.

This, plus later high-speed photographic sequences Muybridge created, such as a famous "movie" of a trotting bison, might conclude his relevance to our story, except that his life in San Francisco kept getting too weird to let us leave him just yet.

In 1872, at age forty-two, he married a twenty-one-year-old divorcée named Flora Shallcross Stone. Three years later, Muybridge happened upon a letter to his young wife from one of her friends, the drama critic Major Harry Larkyns, which made him suspect that Larkyns might be the father of their seven-month-old son, Florado. (His suspicions may not have been too far-fetched; unbeknownst to him, she'd already sent Larkyns a photo of the boy with the caption LITTLE HARRY.)

On October 17, 1875, Muybridge embarked on the six-hour trip from San Francisco to the tiny town of Calistoga, in Napa County, where he tracked down Larkyns. Finding himself face-to-face with the man, Muybridge said, melodramatically, "Good evening, Major. My name is Muybridge, and here's the answer to the letter you sent my wife."

With that, Muybridge shot him in the chest at point-blank range. When Larkyns died that night, Muybridge was arrested, locked in jail, and charged with murder. The story gained continual headlines in the gossipy city to the south, as did the subsequent trial.

This, too, proved high theater. Leland Stanford, the former governor, helped pay for a top-notch defense attorney, who pleaded insanity on behalf of his client and produced a series of longtime Muybridge friends who testified that his personality had become unstable after the stagecoach accident fifteen years earlier. But the photographer undercut the insanity plea by insisting that he had killed his wife's suspected lover deliberately and, indeed, had planned it ahead of time.

The jury didn't know what to make of the man sitting in the defendant's chair, who alternated between vacant, Parkinson's-like detachment and loud emotional outbursts. Was he crazy or not? In the end they rejected the insanity defense but found him not guilty anyway, because they viewed Larkyn's murder as an instance of justifiable homicide.

Muybridge sailed to South America on an extended photo shoot after his acquittal. His wife, Flora, tried to obtain a divorce while he was gone, but her petition was rejected by a judge. Five months after the trial, she fell ill and died at the age of twenty-four. Their son, Florado Helios Muybridge, was placed in an orphanage by Muybridge, who had almost nothing to do with him thereafter. (Later photographs of the fully grown Florado Muybridge show him strongly resembling Muybridge and not Larkyns.) The boy worked all his life as a ranch hand and gardener and in 1944, at the age of seventy, was run over and killed by a car in Sacramento. Muybridge himself continued perfecting stop-action photography using a new shutter design he developed, which featured previously unheard-of speeds of a thousandth of a second. In 1894, after the prolific photographer produced more than one hundred thousand action sequences, he returned to England, where he authored two bestselling books of his photographs, *Animals in Motion* (1899) and *The Human Figure in Motion* (1901). He died at age seventy-four while living with his cousin Catherine Smith.

Muybridge paved the way for stop-action photography and slow-motion cinematography, and others soon followed, until the invisible high-speed world was available to everyone. The Austrian physicist Peter Salcher captured a bullet in flight in 1886, and by the middle of the twentieth century technicians were able to achieve shutter speeds in the microsecond range, a thousand times faster than Muybridge's. One image, taken at a shutter speed of just three-millionths of a second, freezes the spine-chilling opening moments of an atomic bomb blast.

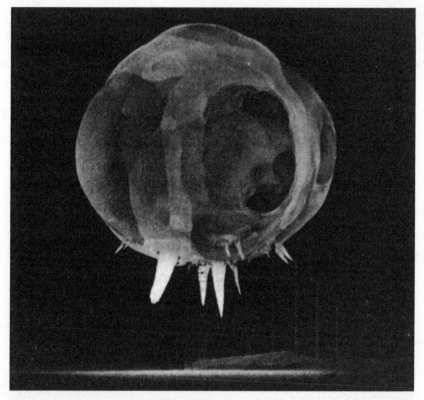

The fastest events require the fastest shutter speeds. A special US government rapatronic camera captured this 1952 atomic bomb test one millisecond after detonation. The exposure was three millionths of a second. Notice the tower still momentarily remains unvaporized. Eerie "rope tricks" of heat, like feelers, spread downward along the unseen mooring cables as they are being vaporized. *(US Air Force 1352nd Photographic Group, Lookout Mountain Station)*

Most of this high-speed action coexists with our everyday lives. We readily perceive the flicker in old silent movies, which run at sixteen frames per second. But we see a steady light in modern cinema's seventy-two frames per second.[4] The human "flicker fusion threshold" is widely regarded as twenty flashes per second. You can experiment yourself if you still have an old strobe light in the attic left over from those psychedelic 1970s parties. Set it to twenty, then twenty-five, then thirty, and see when separate flashes seem to be replaced by solid illumination.

Some say they can perceive those annoying fluorescent light-bulbs flicker, even though the motion must occur at a minimum of three times the usual flicker fusion threshold rate. The ability to detect flickering varies considerably from person to person. Reactions to fast events vary among animals as well. We go to swat a fly, but it leaps out of the way. Faced with a looming swatter, a fly's tiny brain calculates the threat's location, creates an escape plan, and places its legs in an optimal position to hop in the opposite direction—all within a tenth of a second, which happens to be the same duration as an eyeblink.

This maneuver on the part of the fly outwits some fast-reacting animals, such as cats, who can't reliably catch flies. But monkeys can; they seem to live in a faster time and appear to almost effort-lessly pick flies up. Chickens, too, routinely peck flies off a surface.

These events and processes unfold continually around us. In the animal kingdom, the fastest natural motions are the startle reflexes. Such lightning-quick defensive responses often involve an instinct to escape from a sudden threatening situation. They work by using neural electrical mechanisms that completely bypass cerebral pro-cessing and voluntary control. Because the circuits involved are shorter, the durations of startle reflexes are also much shorter than those of voluntary actions. In people, such reflexes can occur in an impressive one-thirtieth of a second. Rats can react even faster, with responses measured at speeds as swift as a thousandth of a second.

Even actions in nonemergency situations can be virtually eye-blink fast. The mandibles of certain ants close around prey in just one-seventeenth of a second and clamp down at eighty miles per hour. Moles, which most of us hardly regard as sprightly, react as soon as their nasal appendages contact a potential food source underground. They strike out in a seventh of a second, about the time it takes to say the second "pa" in the word *papa*.

Among the all-time fastest routine reaction speeds in nature are those observed recently in skipper butterflies. When confronted with a sudden bright light, they respond with a startle reflex in one-sixtieth of a second. Who would imagine that butterflies rank among Earth's quickest-reacting creatures?

Nonbiological processes often proceed faster—much faster—than those in animals and humans. Some chemical reactions happen ten million times faster than an eyeblink, although other reactions, such as iron oxidation (rusting), can take years to unfold.[5] The very speediest? It's nothing exotic: the creation of water from the bonding of oxygen and hydrogen. The protons assume their new positions on a picosecond timescale—a few *trillionths* of a second.

The odd thing is: Why is water a liquid? Composed mostly of the tiniest, lightest atom in the cosmos, it's such a small, feather-weight molecule that it ought to be a gas at room temperature. Other molecules of water's size are gases. Compounds such as methane (CH_4) and stinky hydrogen sulfide (H_2S) closely resemble water in terms of mass and size. Yet they're gases in all natural earthly conditions, even in the Antarctic. Methane boils from liquid to gas at minus 258 degrees Fahrenheit, nearly five hundred degrees lower than water's 212 degrees Fahrenheit. If water behaved "normally," our veins would be filled with vapor, meaning that Earth would be lifeless.

Water's bizarre fluid nature happens because of geometry: when its hydrogen and oxygen atoms connect, they form an odd bend slightly greater than a right angle. This gives the molecule a bit of a polarity, a charge, that lets it weakly connect with other water molecules. It takes a lot more kinetic energy (heat) to break the hydrogen bonds and free the water molecules as a gas. Recent studies show that such hydrogen bonding involves no more than three molecules at a time and takes place in about a trillionth of a second. In a picosecond, the momentarily larger triad structure makes the molecule *act as if it's much bigger than it really is*. Thus water behaves as a liquid at room temperature even though those fragile threesomes come and go hundreds of times each billionth of a second.

We couldn't laugh until we cried—or salivate or bleed or own a brain—without these momentary ultrafast connections.

Picosecond-level activities lie beyond our imagination. Such a time frame requires an example to be even roughly grasped. A photon, then, traveling at the speed of light can circle the earth eight and a half times in one second, but such a light particle would only be able to travel the width of two human hairs in a picosecond, a trillionth of a second.

A trillion seconds? That's thirty-two thousand years. A trillion seconds have not elapsed since we first gained the ability to make fire. A trillion is truly enormous; a trillionth unimaginably tiny.

Even trying to conceive of a mere nanosecond, a paltry *billionth* of a second, can tax our minds, though the word *billion* has become commonplace in today's technologies. We now routinely exploit events that happen in a nanosecond when we use distance-finding devices, for example. The new laser measuring tools, available in any hardware store, shoot pulses of light across a room. A built-in photometer perceives the pinpoint beam reflected from the far wall and calculates the time it takes for the light to make the round-trip. Each nanosecond delay in the light's echo translates to a distance of 11.8 inches. Aim the device next at the adjoining wall and it'll calculate the room's area. Now you know how much paint or carpet to buy. You can toss the old tape measure.

Not all chemical or physical reactions are that fast, of course. Reaction speed depends on the concentration of the substance; whether it's a gas, liquid, or solid; its temperature; and sometimes even bizarre factors such as the room's brightness. Brightness? Since light is energy, it can provide the reacting particles with additional oomph and push them over the edge from slow to fast. Chemists love to demonstrate this by mixing ordinary natural gas with chlorine. When this is done in a dark room, the reaction barely happens at all. It's totally sluggish. Under dim light the reaction speeds up greatly. Do it in direct sunlight, however, and you get an explosion. Light makes the reaction instantaneous.

But the greatest motion influencer is temperature. Turn up the heat, and things speed up. Actually, heat is simply our word for atoms moving, nothing more. It makes sense, then, that the hotter something is the faster its reactions proceed, since electrical bonds between molecules are being broken, electron excitations are occurring, and there is more contact between atoms.

Room-temperature gas molecules typically dart about a bit faster than the speed of sound. Those in your freezer go fifty miles per hour slower. How fast atoms must move to begin an oxidizing or burning process depends on the substance. White phosphorus ignites at just below body temperature; merely holding it is dangerous.

Common combustibles usually require at least four hundred degrees to make their hydrogen combine with oxygen in the air and also in themselves.

What's perilous about most reactions is that they are exothermic. They create heat.

We're thus surrounded by substances poised and ready to burn and to keep burning. Only their atoms' low everyday speed keeps them well behaved. But if some external agent pushes on their atoms to speed them up, the show begins. Once started, they supply enough heat to self-sustain. A simple match is the most common agent of creating such a runaway reaction, a Frankenstein.

The quest for a cheap, portable device that could start fires bore fruit in the eighteenth century and achieved practical success in the nineteenth. Before that, people carried bits of flint or other friction-based, spark-producing materials with them, or else they used a convex lens or concave mirror to concentrate sunlight onto a combustible substance. By the eighteenth century, those who could afford it kept chemical-laced sticks, which were plunged into jars of sulfuric acid to produce violent, dangerous, fire-starting reactions. But in the mid-1800s, white phosphorus's low ignition point proved obvious and irresistible. People could buy "lucifer matches" at any general store. They became so commonplace that lucifers are regularly mentioned in Mark Twain's books, as familiar a cultural touchstone as a smartphone in contemporary literature.

But white phosphorus is a dangerous compound and caused many accidental poisonings; it also became a favorite suicide method. By the turn of the twentieth century it had been largely replaced by red phosphorus and banned outright in many countries. Soon matches were available in two varieties, a situation that still pertains in the present day. *Strike anywhere* types have tips covered in a self-contained, combustible mix of phosphorus sesquisulfide and potassium chlorate. Quickly dragging the match across any rough surface, at a typical speed of six feet per second (four miles per hour), generates enough friction to raise the tip above its self-ignition point of 325 degrees Fahrenheit. Easy.

Sometimes too easy. Matches have never been allowed on planes or ships. The alternative is *safety matches*, which require contact between the matchbook's scratchy surface, which contains a bit of red phosphorus and ground glass or another type of roughener, and the match head, which is about 50 percent potassium chlorate, famous for readily bursting into flame and releasing oxygen. The match head also contains a little antimony trisulfide, a safety component because it needs the combustion heat from the others to ignite. This witches' brew requires a higher friction temperature of 450 degrees or so.

Once ignited, the match's fire quickly reaches a temperature between 1,112 degrees Fahrenheit and 1,472 degrees Fahrenheit; the uppermost part of the flame is the hottest. Its molecules move so fast they can easily jostle those in other substances, which themselves soon reach a speed that lets them initiate their own burn reaction. Frankenstein has come to life. The speed needed to begin a self-sustaining "burning" event depends on the substance.

Fire—important enough to qualify as one of Aristotle's elements—is a motion exhibit in several simultaneous ways. Flames lick the air and dance in a thousand mesmerizing patterns, while the unseen choreography is just as intriguing.

Paper "catches" easily. Its famous ignition temperature inspired the title of the 1953 Ray Bradbury novel, *Fahrenheit 451*, in which "firemen" went about burning books. Catchy, but in reality, different types and thicknesses of paper have distinct ignition points that vary from 424 degrees to 475 degrees. Real science is often less succinct than its fictionalized analogues. (Bradbury, who died in 2012, knew this, of course. He also knew that the title *Somewhere Between 424 and 475 Fahrenheit* would not have been as memorable.)

Coal ignites reluctantly, with a very high ignition point of 842 degrees. Kerosene is much more eager to go, at 444 degrees. Gasoline will ignite at 495 degrees, alcohol at 689, and hydrogen at 752. But there are nuances, especially with flammable liquids. When atomized as a spray, home heating oil burns brilliantly. But a two-foot-deep pool of it that has leaked into a basement is unlikely to ignite even if a lit match is tossed into it. Similarly, you can squirt Pam cooking spray into any candle flame and it'll instantly whoosh as a brilliant blaze. But even if heated to its self-ignition point of 644 degrees Fahrenheit, cooking oil may not burn unless it's atomized, which envelops it with needed oxygen. That temperature is slightly higher than what's normally achieved in an oven, which is why oils aren't frighteningly ablaze whenever you check on a baking eggplant Parmesan.

Eerily enough, cool molecules can sometimes move faster and faster on their own until before you know it your house has burned to the ground. No spark or flame is needed. This *spontaneous combustion* begins when something with a relatively low ignition temperature, such as rags, straw, or even wheat flour, remains in contact with moisture and air. These substances supply oxygen and allow bacterial growth that encourages fermentation. This in turn generates heat, as anyone with a compost pile or a stack of rotting hay can confirm. If the heat is confined and unable to dissipate (e.g., if oily rags are crammed in a pail or buried in a pile of hay, which is a good thermal insulator on its own), the temperature rises, eventually exceeding the ignition point. The result is a thermal runaway.

Pyrophoric substances are those whose molecules can explosively increase their speed with very little provocation. They're wound up and raring to go. They burst into flame at room temperature or less. Sodium is a famous example. Its autoignition temperature is exceeded almost everywhere in everyday life, and it will undergo a violent reaction when in contact with water or even moisture.

There have been grain elevator fires in which no culprit spark triggered the explosion. Items as seemingly innocuous as corn have blown up when moisture was allowed to accumulate. Among the substances most susceptible to spontaneous combustion are pistachio nuts. You can't make this stuff up.

The point is, it's all motion. Heat is motion. Atoms' motion is heat. When you run a fever you might complain that your temperature is 102. But you could just as well tell the doctor, "I feel awful. My body's molecules are moving three miles per hour faster than normal."

Then he'd hand you some aspirin, saying, "Here. This will slow them down."

You can also get a rough idea of atomic speed and temperature by a substance's color when heated. It's wonderfully simple. It doesn't matter what it is. Cast iron, copper, the tungsten in lightbulbs—when a noncombustible object starts glowing, the color of that light is an accurate guide to its temperature.

214

TRANSLATING GLOW COLOR
INTO TEMPERATURE

If a substance glows a dull red, just barely visible in the dark, it's 752°F.
A red heat visible in subdued lighting means 885°.
If the red can be seen in daylight, it's 975°.
If it's visible as red in direct sunlight, it's 1,077°.
If it's cherry red, it's 1,650°.
Orange indicates 2,012°.
Yellow means 2,370°.
White indicates a temperature of 2,730° or higher.

Upon attaining the color white, you may well have trespassed beyond the object's melting point. In any case, white is the end of the line. In theory, an even hotter substance would glow blue— just as blue stars are the hottest in the universe—but by then all earthly materials would have melted if not boiled into gas.[6] Finding a substance that would remain solid when white hot was what created so many headaches for Edison as he struggled to perfect his electric light. He finally found tungsten, the element with the second-highest melting point, which stays solid until it's a whopping 6,170 degrees Fahrenheit. This was vital: a thin lightbulb filament is asked to remain at an amazingly high 4,500 degrees for hours or even days at a time; that's about *twice as hot* as melting steel. (Carbon has a slightly higher melting point but is too brittle to be practical as a filament.) The incandescent bulb's fantastic heat would ultimately prove its own undoing: it is now being replaced by LEDs or fluorescents or banned outright. The complaint is that incandescent bulbs utilize most of their electricity for the production of heat rather than light.

Aluminum melts at a mere 1,220 degrees Fahrenheit. For copper it's 1,976; for gold, 1,945. Unlike common steel, which melts at

2,500 degrees or so, these metals turn liquid at lower temperatures than their glow-white point. That's why you'll never see a white-hot aluminum ingot, just as you never see a red-hot chunk of solid tin or lead, which melt before they can glow in any way.

As for their atoms' speeds, the main motion in a solid is a very small amplitude vibration around its equilibrium position. These vibrations grow larger and more frantic with increasing temperature until the melting point frees them. But only atoms in gases break the sound barrier.

Unbeknownst to Eadweard Muybridge, ultrafast rhythms that rule virtually every aspect of our lives are not rare phenoms, nor could they ever be captured by his or anyone else's camera, then or now.

These astonishing discoveries began in the late nineteenth century, when physicists started finding strange small-scale vibrations. Probably the coolest and most useful example is the piezoelectric effect, discovered by the Curie brothers, Jacques and Pierre, in 1880. They found that many kinds of crystal (they liked to work with quartz) naturally vibrate tens of thousands of times a second if a little electricity is applied to them. And it works the opposite way, too. If a crystal is compressed or distorted or struck so that it vibrates, it briefly *produces* electricity. It's a two-way street.

A flood of technology arose from this. Breakthroughs occurring between 1921 and 1927, mostly at Bell Labs, resulted in the creation of a superaccurate clock that relied on those quartz vibrations. Vacuum tubes and other bulky components confined the early devices to laboratories, where they kept America's official time to a new level of precision on behalf of the National Bureau of Standards (now the National Institute of Standards and Technology) for thirty years, until atomic clocks took over in the 1960s.

Cheap semiconductor technology enabled manufacturers to mass-produce quartz watches beginning in 1969, when they replaced mechanical spring watches and made it possible for everyone to have a personal timepiece accurate within one second per month. Your watch's quartz crystal is shaped to naturally vibrate 32,768 times a second. This is, conveniently, a power of two (it's two multiplied by itself fifteen times over), which lets digital circuitry easily convert it into whole seconds.

Pulsating crystals are now in every home. For example, you may own one of those barbecue grill lighters—the ones with an annoyingly hard trigger. Pulling the trigger strikes a crystal, which piezoelectrically creates a momentary high voltage, thus producing a quick spark. No battery is ever needed. Indeed, gas stoves now use vibrating crystals to create the spark that ignites the gas. If you hear repeated "snaps" whenever you turn it on, that's what's happening.

Thirty-two thousand vibrations a second may seem fast. But it turns out it's not just crystals that undulate. *Absolutely everything vibrates.* The molecules that make up every substance around us display complex atomic harmonic oscillations.

We may imagine that a simple common compound such as water, made of two hydrogen atoms bonded electrically to an oxygen atom, has a rigid structure. Not so. The atoms stretch away from each other a bit and then snap back as if on a rubber band. At the same time they twist around and then return to shape. They also rock back and forth like a metronome. Each of these repetitive atom motions—twisting, stretching, rocking, bending, and wagging—has its own precise period that is somewhere between one trillion and one hundred trillion times per second. You'd think this shaking would dampen out and stop. It never does.

Meanwhile, light itself consists of waves of magnetism and electricity whose pulsation rates depend on the color. Green light's waves, for example, pulse 550 trillion times per second. These vibrations are not just extraordinarily regular. They have in-your-face consequences.

To give one example, an automobile parked in sunlight heats up because the pulsation rate of the infrared waves that are inside it coincidentally match the atomic vibration rate of the car's glass. This creates a chaotic boundary, blocking the heat from escaping through the windows the way light does. Instead, light gets in, but the heat it creates can't get back out. It makes the car's interior very uncomfortable when you step inside. People have been arrested for leaving pets and children in such parked cars. The charges probably didn't specify that the suspects "ignored the lethal perils of ultra-fast vibrations," but that's what it amounts to.

As another example, consider the chrome that adorns motorcycles and makes the exposed metal parts of cars look so shiny. This happens because the outer electrons of the element chromium absorb and then reradiate the photons of light that hit them. But the light doesn't get too much farther. That metal's inner electrons are held so tightly in their orbits that they have too little flexibility to vibrate and give off light. The end result is that sunlight striking chrome and most other metals isn't fully absorbed, nor does the light pass through. It's neither transparent nor dull but something else: gleamy.

So we're immersed in more than mere animation. Nature doesn't just go wild with countless pulsations producing powerful everyday experiences. It also never tires of repeating itself within time frames of the tiniest fractions of a second—or in milliseconds, whole seconds, minutes, years, centuries, millennia, you name it. Ours is a shimmering, vibrating universe on multiple levels. These patterns, which interact with each other, influence everything—even if we're unaware of virtually all of them.

CHAPTER 15: *Barriers of Light and Sound*

A Thirty-Century Quest That Began with Thunder

As fast as it can go, the speed of light, you know
Twelve million miles a minute...
— ERIC IDLE AND TREVOR JONES, "THE GALAXY SONG" (1983)

The sound barrier. The speed of light.

These classic entities created endless mind torture for Homo bewilderus. We who are utterly dependent on sight and sound learned early on that nature performs its symphonies prestissimo. Even the people most associated with sight and sound gained renown, such as the sound-barrier-breaking Chuck Yeager and light's maestro Albert Einstein, who symbolized its speed with a lowercase c in his famous $E = mc^2$ equation.

The head-scratching began far before those twentieth-century celebrities captured the limelight. It may have all started with thunderstorms. Here is nature's *only* exhibition of simultaneous brilliant light and deafening sound. It always attracts attention. As in the days of Aristotle, a lightning bolt can be temporarily blinding. A thunderclap can rattle dishes. The birds outside grow ominously silent.

These days, at least, we regard thunderstorms scientifically. When the flash comes we think "electricity" and console ourselves with the fact that fewer than one hundred Americans a year are killed by lightning. Odds are it won't be you. If you are female rather than male (men are struck five times more often), live in any state except Florida (the most lightning-prone state by far), and neither fish nor play golf (the most lightning-attracting activities), you can maybe even let yourself enjoy the violence.[1]

Back when the Parthenon was built, when there were fewer lightning-prone golf courses, thunderstorms were always exhibitions of—you knew this was coming—godly power. The word *thunder* comes from the name of the old Norse god Thor, the hammer-wielding deity who also gave us the word *Thursday*.

Yet he scarcely had a monopoly. The Bible makes many references to lightning being dispensed by Jehovah. The first occurs in Exodus 9:23: "And Moses stretched forth his rod toward heaven: and the Lord sent thunder and hail, and the fire ran along upon the ground."

This intimidating amusement was also a specialty of a wide pantheon of Roman and Greek deities. The ultimate thunder hurlers were the German god Donar and the Greek god Zeus, whom the Romans called Jupiter. When you traveled east, if you'd managed to evade the wrath of the European gods, you'd instead get smitten by the Slavic god Perkunis and then the Indian god Indra.

The lightning bolt itself was often depicted as a spear. In Roman times, whatever it hit was considered sacred. Sometimes, where melted glassy sand marked a strike, the place was so revered it was fenced off. The authorities didn't quite charge admission, but people killed by lightning were conveniently buried right in that consecrated place rather than carted off to a burial ground. In African and South American cultures, the mythical giant thunderbird was often fingered as the cause of storms.

During the classical Greek period, when science and observation flourished, the importance of vision and hearing generated widespread speculation about how sounds and images might go from point A to point B. And while the hypothesizing is certainly not over, the basic, baffling, early puzzles have now metamorphosed into a continuing modern fountain of gee-whiz science.[2]

The century during which much of the Old Testament was penned witnessed the first nonreligious ideas about visual and auditory fury, proposed by the Greek thinker Thales (ca. 620–546 BCE) and his followers Anaximander (ca. 611–547 BCE) and Anaximenes (ca. 585–528 BCE). The three get merit points for stepping away from the Zeus-hurling-spears business, even if their conclusions were wrong. They each wrote that thunder is wind smashing through clouds, a process they believed kindled the flame of lightning. Thus thunder came first, a conclusion embraced, curiously, for the next two thousand years.

Not that there weren't any dissenters. Anaxagoras (ca. 499–427 BCE) said that fire somehow flashed first, only to be doused by the clouds' rain. Thunder, he believed, was the sound of the lightning being violently extinguished.

Aristotle, his brain crammed with intricate beliefs about everything, entered the fray in his ca. 334 BCE collection of essays called *Meteorologica*. There he sided with Thales. He wrote that thunder is the sound of trapped air in clouds slamming into other clouds, and added, "Lightning is produced after the impact and so later than thunder, but it appears to us to precede it because we see the flash before we hear the noise."

It wasn't entirely malarkey. Here was the very first known statement that light *moves* faster than sound. This may seem like a groundbreaking notion, one more bit of evidence that Aristotle belonged in an accelerated classroom. In actuality, determining the *relative* speeds of sound and light never required a genius IQ. Echoes within great halls and canyons had always suggested that sound was a slowpoke.

The colliding-clouds explanation for thunderstorms remained popular for century after century. Around the middle of the first century BCE, the Roman poet Lucretius, in his *On the Nature of Things*, wrote of thunder:

> *The winds are battling. For never a sound there comes*
> *From out the serene regions of the sky;*
> *But wheresoever in a host more dense*
> *The clouds foregather, thence more often comes*
> *A crash with mighty rumbling.*

Why didn't the 100 percent more correct lightning-first idea catch on? Probably because in that pre-gunpowder era it would have been a unique event, totally without precedent.[3] A light never caused a sound, especially in the heavens. The sun and moon are of course silent. So are common exploding meteors and fireballs. The aurora is mute as well. Even in the biological realm of fireflies and phosphorescent marine creatures, the glows unfold in silence.

Meanwhile, some of the early Greeks went beyond storms to probe the very nature of sound. Pythagoras (ca. 570–495 BCE) wondered why some combinations of musical notes sounded more lovely than others. He made a startling discovery. Experimenting with vibrating strings of various lengths, he found that when they had whole-number ratios to each other the resultant combination was always pleasant and harmonious. For example, if a plucked string produces the musical note A, a string twice as long will also create an A, but an octave lower, corresponding to a numerical ratio of 2:1. The notes in between are produced by plucking strings that have string-length ratios such as 8:5, 3:2, 4:3, and so on. Later, Aristotle correctly wrote that sound is nothing more than the expansion and contraction of air produced by air's proximity to pulsating or oscillating objects, such as strings, rustling leaves, vocal cords, and the vibrating bronze of a bell.

And that's where things stood, with no further acoustic progress as century bells tolled and tolled again. Sound remained a mysterious subject right into the dawn of the scientific revolution. In the early 1600s, Shakespeare had King Lear ask (in act 3, scene 4): "What is the cause of thunder?" and there was no answer. At around that same time, in 1637, René Descartes—who wrote cogently about optics and vision and sound propagation—nonetheless maintained that colliding clouds create thunder, wrongly echoing the Greeks two thousand years earlier.

But things were starting to change. We celebrated Galileo in our exploration of gravity and falling bodies, but the great man's observations about sound were spot-on, too. Early in the seventeenth century he wrote, "Waves are produced by the vibrations of a sonorous body, which spread through the air, bringing to the tympanum of the ear a stimulus which the mind interprets as sound."

Notice that this answers the age-old "If a tree falls in the forest" conundrum. Nowadays the great majority of people will opine that a falling tree makes a sound even if no one is around to hear it. Not so, suggested Galileo. Rather, the crashing oak creates complex puffs of pressurized air—actually a consortium of subpuffs, or thousands of very brief individual air pressure changes—that spread outward. These brief tiny breezes have no inherent sound. Instead, these silent puffs set the ear's tympanic membrane vibrating in a very detailed way, with slow pulsations and fast ones superimposed on each other—which, Galileo observed, "the mind interprets as sound." Galileo was bang-on correct. He advocated something rarely heard before the advent of quantum mechanics: the importance of the observer. We know now that nature and the conscious observer are correlative. They go together. Sound requires the presence of both.

Galileo thus basically defined sound as pressure waves. Rapid, intricate puffs of wind. A disturbance in air or another substance. Later researchers found that a human can perceive a noise—that is, her eardrum will be stimulated enough to vibrate—if the ambient air pressure momentarily fluctuates by just one part in a billion. And, moreover, she'll hear a sound only if those air pulses don't repeat themselves more than twenty thousand times or fewer than fifteen times in a second. Those are the parameters for human aural response, which make nerves in the eardrum send electrical signals to the brain. Outside that range, the quick tiny breezes unfold in silence.

After Galileo the science of sound advanced quickly, the revelations ever more amazing. Thunderstorms surrendered their cacophonous secrets as well. Benjamin Franklin famously discovered lightning's true nature on June 10, 1752, during a dangerous kite-flying experiment that might easily have killed him. He correctly concluded that lightning produces thunder and not the other way around. After all, he had already created sparks in his laboratory, and each was always accompanied by an audible *snap*. Franklin

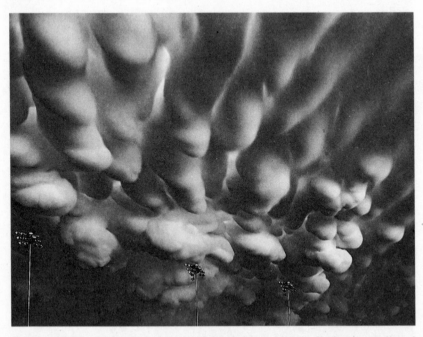

Mammatus clouds contain such violently high winds that all planes, small and large, give them a wide berth. *(Jorn C. Olsen)*

wrote, rhetorically, "How loud must be the crack of 10,000 acres of electrified cloud?" (He was no mere dabbler. He'd been obsessed with uncovering the secrets of electricity for more than a decade. It was he who coined the words *electrician, conductor,* and *battery*.)

Joseph-Nicolas Delisle (1688–1768) went even further. This French astronomer, who made a killing building an observatory and then setting up Russia's astronomy program for Peter the Great, started studying thunderstorms at the age of fifty. He decided that while lightning can be seen at great distances, even from more than one hundred miles away, thunder is generally inaudible if the bolt lies farther than a mere fifteen miles away. Even in our time, people still mistakenly attribute silent flashes to "heat lightning"—a nonexistent phenomenon—unaware that these are just distant thunderstorms whose sound waves have already fully dissipated.

Back then, on the rare occasions when a lightning bolt made ground contact and left a scar, a precise distance from the observer could be determined. Using that distance and the previously timed delay between the flash and the thunderclap, "natural philosophers" had no trouble pegging sound at 768 miles per hour. But sound's speed became *popularly* famous because of a single intriguing concept: a sound *barrier*. This demands attention because it seems like a challenge. There's no smell barrier or light barrier. Why sound alone?

This idea arose in the modern aviation era, before which no man-made object except for bullwhip tips and bullets had ever gone that fast. The barrier is a problem because air's pressure waves accumulate on any object approaching sound speed. That's what produces a sonic boom. These dense air compressions created squirrelly control issues on jet aircraft in the early 1950s as pilots attempted to reach that speed. As for the drawn-out rolling of thunder, nineteenth-century scientists rightly attributed it to the sound from nearer sections of the lightning bolt arriving ahead of the rest. Because electrical strokes can easily exceed a mile in length, the incoming sound from its various segments can maintain the rumbling for more than five seconds.

Yet even into the twentieth century, when thunderclouds started to share the celestial stage with flying machines, no one knew how lightning creates thunder. There were three excellent theories, and all seemed plausible. Go ahead, try your luck. Which of these would you vote for?

The steam theory of 1903 said that a lighting stroke suddenly vaporizes all a cloud's water along its route. This high-pressure steam violently expands, producing the thunder sound, just as it does inside an exploding locomotive boiler.

Another theory, from 1870—at the frenzied height of chemistry advances—held that the electricity in the lightning bolt, like an electrode in a beaker of water, disassociates the cloud's water into separate hydrogen and oxygen atoms. When these quickly recombine, the result is a big boom. After all, a mixture of those elements always explodes if a spark is at hand. That's exactly what happened during the *Challenger* shuttle disaster of 1986.

The third idea was published in *Scientific American* in 1888. One M. Hirn advanced the theory that "the sound which is known as thunder is due simply to the fact that the air traversed by an electric spark, that is, a flash of lightning, is suddenly raised to a very high temperature, and has its volume considerably increased. The column of gas thus suddenly heated and expanded is sometimes several miles long and...it follows that the noise bursts forth at once from the whole column, though for an observer in any one place, it commences where lightning is at least distance."

And this last hypothesis—after decades of debate—was accepted by the scientific community. Thunder is explosively expanding air.

All motion. Grand motion. Of an electrical arc on steroids. And supersonic gas expansion. Then pressure waves racing at the speed of sound at right angles to the bolt.

The details came later. Ah, but what details! A lightning bolt is a 55,000-degree sizzle created in ten milliseconds—far hotter than the sun's comparatively lukewarm 11,000-degree surface. By comparison, steel, that symbol of strength, turns to liquid at "just" 2,500 degrees Fahrenheit. The crazy heat of the lightning bolt breaks atoms into pieces, leaving a wildly expanding plasma that creates a pressure ten times greater than the surrounding air. No wonder such storms don't tiptoe quietly.

The frenetically expanding gas generates a wide spectrum of sound. But the high notes dissipate right away. In a matter of feet they're finished. Treble waves pulse quickly and just can't be sustained. That's why, when a teenager's car with its pounding radio passes by, you only hear the thudding bass. The music's highs don't even survive the thirty-foot distance to your curbside ears. It also explains why foghorns are designed to produce only low tones. These travel far. A high pitch would be futile.

The thunder's sound therefore grows deeper the farther it travels. Thus a lightning bolt's sound track reveals the flash's location in three ways: by its loudness, by how sharply defined (as opposed to muddy) the noise is, and by the pitch. If you've ever almost been struck—in which case the sound and flash would be simultaneous—you've heard a much more balanced musical composition, with lots of sharp, crackly high tones. This actually happened right outside my home office just after this book was completed, in time to be included here. The flash and earsplitting crack were perfectly synchronous, as if nature were saying, "You want to relate the experience firsthand? Okay, here you go!" And indeed, the deafening explosion's pitch wasn't in the bass range at all. When thunder's rumble is low and indistinct, it's always more than two miles away.

For real precision, however, the old rule still stands. You count the seconds between the flash and the first sound. Each second means the bolt is another 1,100 feet away. Five seconds pegs the lightning at a mile — almost on the nose.

Everyone can readily perceive a light-sound gap of one-eighth of a second, which corresponds to 150 feet. So when the flash and the boom seem truly simultaneous, the bolt is nearer than half a short city block, a literal stone's throw. A close call indeed.

When we speak of the speed of sound we normally mean the speed at which noise moves through air. But sound travels differently through various substances, even through other gases. We acquire a munchkin voice after inhaling a little helium from a party balloon because the sound of our voices speeds up to a wild 3,200 feet per second through helium, or three times faster than it travels through normal air. Sound zooms even faster through liquids and nonporous solids. It travels 4.3 times more quickly through water and fifteen times more quickly through the steel in railroad tracks than it does through air.[4]

Meanwhile, the lightning bolt, the cause of all this sound and fury, travels to the observer at the speed of light — *a million times faster than sound*. Essentially it arrives instantaneously. To be precise, a mile-away bolt is seen 0.000005 seconds after it occurs.

For millennia, light was assumed to enjoy a speed so fast it couldn't be measured. It was perceived to arrive at distant locations at the same instant of its creation. When the truth finally leaked out in the seventeenth century, light became better understood in some ways but also grew curioser and curioser to those who studied it. Even today, few among us who are not science teachers can come right out and state what light really is. It is easier to rattle off its velocity — 186,282.4 miles per second — than what it's made of. That's true whether we think of light as a particle or a wave.

A wave may seem like something that moves in a visually

obvious fashion, but it never involves the forward movement of anything. No actual substance is advancing. As an ocean wave passes over a particular piece of embedded kelp, the vegetation just bobs up and down. So as we saw in chapter 13, while a wave has a forward motion, the water it's made of does not.

The same is true of sound. A friend shouts hello from across the atrium of a mall. But nothing has traveled from him to you. He merely created a disturbance in the air in front of his mouth, where one molecule jostled the next and so on until the molecule adjacent to your own eardrum caused that membrane to vibrate. No physical object, not a single atom, traveled even an inch.

A classic demonstration of this involves a long rope hanging from the top of a flagpole or scaffolding. By giving the bottom a snap, we create a beautiful wave pattern that fluidly scampers all the way up. It looks like a sine wave performing a lively vertical motion, but in reality each portion of the rope is merely waving back and forth.

So with light, the question right from the beginning became: What exactly is moving?

The ancient Greeks believed light was a beam that traveled outward from the eye. But other early thinkers thought vision was an interplay between this eye beam and something emitted by light sources such as the sun. The Greek who came closest to the truth was Lucretius. In his *On the Nature of Things,* he wrote, "The light and heat of the sun are composed of minute atoms which, when they are shoved off, lose no time in shooting right across the interspace of air in the direction imparted by the shove."

Lucretius's view of light as particles—ultimately supported by Isaac Newton—included that profound "lose no time" clause, which meant simultaneity. In any case, light remained popularly regarded as just an eye phenomenon for centuries to come.

It took a full millennium for anything to change. The next true breakthroughs came from Alhazen, whom we've already met for his accurate appraisal of the atmosphere. Around 1020 CE he said that vision results solely from light entering the eye; nothing emanates from the eye itself. His popular pinhole camera lent weight to his arguments. But Alhazen went much further. Elaborating brilliantly, he said that light consists of streams of tiny, straight-moving particles that come from the sun and are reflected by objects. He insisted that light travels at a fast but *finite* speed. Refraction—the bending of light, as when the setting sun looks distorted—is caused, he said, by *light slowing down* as it passes through substances of progressively greater densities, such as the thick air near the horizon.

Alhazen was absolutely right. His on-the-nose conclusions were six centuries ahead of anyone else's. For example, Kepler, in 1604, made astute observations about light but nonetheless believed that it moved infinitely fast, and a generation later Descartes seconded this wrong view. Worse, Descartes kept publishing arguments for infinite speed and announced that he would "stake his reputation on it."

The day arrived when certainty became an illusion. Yet these great men shouldn't be mocked in hindsight: infinite speed was a very far-out concept. We can all imagine an extremely fast entity; everyone knew that at the very least, light had to be off-the-charts speedy. But something arriving as soon as it departs? Requiring no time at all? This would make light unique in all of nature. (Turns out quantum phenomena go infinitely fast, as we'll see in the penultimate chapter.)

Meanwhile the "What is it?" debate raged on. It grew heated, almost taking on the quality of a food fight. In the late seventeenth century Newton joined Kepler in arguing that light is a stream of particles, while the likes of Robert Hooke, Christiaan Huygens, and, soon, Leonhard Euler insisted it's a wave. Of course, if it is, what is it a wave of? These Renaissance scientists were thus forced to believe that space is filled with a plenum (later called an ether) because there had to be a substance that actually waves.

One obvious fact managed to sway many in favor of Newton's particle idea. When light from an angularly small or distant source such as the sun passes a sharp edge, such as the wall of a house, it casts a sharp-edged shadow on nearby objects. That's what particles moving in a straight line should do. If, instead, light were made of waves, it ought to spread out—*diffract*—as ripples do and as sea waves do when passing a jetty. Sharp-edged shadows, added to Newton's reputation as a genius, made the wave proponents seem like nut jobs.

Meanwhile the finite-versus-infinite uproar finally ended when the Danish observer Ole Rømer determined light's speed in 1676. His idea was simple. Anyone with a small telescope could watch Jupiter's four large moons speed up and slow down as they whirled around the giant planet during their 399-day cycle. Meaning that for about half a year they moved faster than they did during the other half. This made easily observed events, such as each moon passing in front of Jupiter, occur up to fifteen minutes "early" or "late" compared to its average orbital speed. The moons zoomed faster whenever Earth approached Jupiter. Conversely, the moons grew strangely sluggish when Earth chugged away.

Something went *boing* in Rømer's mind, and he dropped his pastry in midbite. Each image of Jupiter's animated environment had to travel farther to reach us whenever Earth was flying away from it! At such times our two worlds grew nineteen miles farther apart each second. Every "frame" in the movie, as we might visualize it today, needed to go farther than the last, and this took a bit of time. *Of course* the scenes would then seem to run in slow motion. The delay proved that light isn't infinitely fast.

The great Dane calculated light's speed as 140,000 miles per second. Since the correct speed of 186,282 miles per second was not determined for another two centuries, Rømer did well to underestimate it by a mere 25 percent. Indeed, there was no way to do better without knowing Jupiter's true distance from Earth, which wasn't reasonably revealed until another three generations had come and gone.

This is not the place to recount the full, fascinating story of geniuses such as Augustin-Jean Fresnel, Siméon-Denis Poisson, Michael Faraday, James Clerk Maxwell, Max Planck, and Albert Einstein, who each made brilliant breakthroughs in understanding light. Or the quantum-mechanics gang—Satyendra Nath Bose, Niels Bohr, Louis de Broglie, Werner Heisenberg, and Erwin Schrödinger—who made it clearer but stranger. Our mandate is solely that of speed and motion.

Still, a couple of minutes can clarify what, exactly, is moving.

The particles-versus-waves controversy? As if some wise King Solomon ruled nature, *everyone* was soon declared right.[5] Scottish physicist and mathematician James Clerk Maxwell showed that light is a self-sustaining wave of magnetism and an electric pulse positioned at right angles to it. They go together, creating each other in a mutually nurturing way. Light is thus justifiably called an *electromagnetic* phenomenon. Unlike sound, light is not a disturbance in some medium. Light exists on its own. It's quite content to fly through the vacuum of space.

The "electro" part of the word, helpfully, sounds like "electron," the first subatomic particle discovered, in 1899. That's no coincidence. Turns out light is born one way and one way only. If an atom gets struck by energy, this excites its electrons, which give a figurative yelp and jump to an orbit farther from the nucleus. They don't like to be there. So in a fraction of a second they fall back to a closer orbit. As they do so, the atom loses a bit of energy. This is instantly converted into a bit of light, which materializes out of the emptiness like magic and then rushes away at its famous speed.

That's the only way light is ever born. Out of seeming nothingness, whenever an electron moves closer to its atom's center. Simple, really. But ask all your friends how light is created and you'll get blank stares.

So light is a wave of electricity and magnetism. At least, that's the best way to visualize it as it's traveling. But when it starts off and also when it collides with something, it instead acts as a tiny bullet, a massless particle, a photon. Nowadays one may call it a photon or a wave and be equally correct. However you slice it, there's a lot of light in the cosmos — one billion photons of light for every subatomic particle of matter.

Anyway, the quantum guys showed that solid objects such as electrons can behave as waves of energy, too. When an observer uses an experimental apparatus to determine the location of the photons or the electrons in an atom, the photons and electrons will always behave as particles and do things only particles can do, such as pass through one little hole or another but not both at once. But when no one's measuring where exactly each is situated, they behave as waves that blurrily pass through both holes simultaneously to create an interference pattern on a detector on the other side — which only waves can do.

Thus the observer plays a critical role in what he or she sees. Most physicists now think that a human consciousness is required to make an electron's "wave function" collapse so that it occupies a particular place as a particle. Otherwise it's just an indefinite probability item with neither location nor motion.

But does an electron's wave function collapse and turn into an actual particle if a *cat* is watching? Would light always be waves and *never* discrete photons if no one were around? Our best answers are "Who knows?" and "Yes," respectively, but obviously this whole business is Wonderland-strange.

Light's speed is counterintuitive, too. In a vacuum a photon always travels 186,282 miles in a second. Its reputation as a constant is well deserved, but only when it's flying through the emptiness of space. When passing through denser transparent media, such as water or glass, photons seem to slow down. *Seem?* Well, do they slow or don't they?

You decide.

Light moves at just two-thirds its regular velocity, at "just" 120,000 miles a second, through glass and at 140,000 miles per second through water. This speed change is not at all subtle or cerebral. It makes fish appear in illusory locations in a fishbowl and causes a spoon to appear bent in a half glass of water. The density of glass allows a bottle of soda pop to seem to hold much more than it really does.

But look more closely: photons are colliding with the material's atoms, getting absorbed, and then new photons are created to continue on. The images you see through your window are made of different photons from the ones that originally struck the far side of the glass. Since the process of absorption and reemission takes a tiny bit of time, light requires longer to pass through a window than it does through air. But between the glass's atoms each photon really still zooms at its famous, superfast, constant speed.

Each of light's colors traverses clear materials at its own speed. This variance makes their paths diverge. Alhazen knew this a millennium ago.

The bending of each color's path as a result of its separate speed is called *refraction*. We see this when sunlight hits a prism and its component colors get bent into a gorgeous, spread-out spectrum on a wall. Newton explained this by creating a slam-dunk apparatus that silenced all disbelievers. Before him, everyone thought glass merely introduced a distortion that cooked up the colors. Newton won them over by having the colors hit a second, reversed prism that rebent and recombined them. White light emerged. If glass created color as a result of distortion, then Newton's double prism would have produced *more* distortion. Instead he proved that when we see white it's our eye's response to the mixture of all colors.

White light is a rainbow in a blender.[6]

But in a vacuum, all colors fly at the same speed. It is the fastest speed anything can move, and, indeed, nothing with any weight

234

can quite attain it. Light travels a foot of distance in a nanosecond, a billionth of a second. So when we see something ten feet away, we view its image not as it is now but as it was ten nanoseconds ago. We always view the past.

We observe the sun as it was eight and a half minutes ago. If it should blow up at this moment, we'd be given a short reprieve from this disconcerting news. We see the stars as they were years or centuries ago. Galaxies as they were millions or billions of years ago.[7]

We use light's fast but finite speed for some of our favorite technologies. A GPS satellite contains an atomic clock that sends out time signals. Your car's GPS receiver recognizes that it's the wrong time. Wrong because the signal, traveling at light speed, required one-twentieth of a second to arrive at your Camry from the satellite eleven thousand miles overhead. Your GPS, which knows the right time, then instantly calculates how far away that satellite must be for the signal to have been delayed for exactly that amount of time. It does this with three or four satellites and thus triangulates what your position must be. All by using the known speed of light.

But light's constancy is *too* perfect. It doesn't make sense. If you fly toward the sun while measuring its incoming photons, the act of colliding with them should, logically, make them hit you faster, the result of your own speed added to theirs. Or if you race away from a lightbulb at nearly the speed of light, you'd think each photon would barely catch up to you and that its measured speed of arrival would be slower. Not so. In all cases, light always strikes you going 186,282 miles a second.[8]

While orbiting the sun, Earth whooshes toward the orange star Antares at nineteen miles a second in August, and we rush directly away from it each February. But do its photons arrive here at different seasonal speeds? Not in the least. It's as if we're standing still.

So light's constant speed is worse than counterintuitive. It's screwy. And amazing.[9]

Turns out distances shrink and time alters its rate of passage in ways we don't even notice, all so that we perceive light going the same speed no matter what. There's something about that 186,282 miles per second business. Light somehow occupies a more fundamental reality than time and space and everything else we used to think was unalterable.

To conclude our story about trying to find light's true speed—that migraine-inducing quest that endured for so many centuries—we last applauded Ole Rømer's Jovian moons method, which delivered a figure just 25 percent off. But wouldn't it be marvelous and satisfying, and win accolades from colleagues, if one could measure it here on Earth? Newton and his contemporaries tried standing on hilltops with bright lanterns blocked by a quick-release shade. They had a confederate stationed on another hill a few miles away, instructed to unveil his own lantern the moment he saw the first one's light. The first person should simply be able to time the interval between opening his own shade and seeing his partner's returning beam. But when this was done, the delay was always the split-second interval expected from human reflexes. (Turns out the reflection from a mirror positioned even twenty miles away would actually yield a delay of only one five-thousandth of a second.)

The fog finally lifted in 1850, when Léon Foucault improved on an apparatus invented by another Frenchman and nailed light speed at last. The idea was to bounce light from a fast-spinning polygonal mirror onto a flat mirror and back again. During the photon's brief flight through the air, the rotating mirror's angle changes, and the beam reflects in a slightly different direction, readable through a microscopelike device calibrated with fine markings. Knowing the mirror's spin rate and hence its angle's change, and knowing the total distance the beam travels, which in Foucault's case was twenty miles, one can measure light's speed with certainty. Albert Michelson refined the method seventy-five years later, and that's when light speed was confidently known with a margin of error of just two miles a second.

By the time I was performing this demonstration in college, the whole apparatus fit in a lab room, the margin of error was less than one mile a second, and nowadays lasers make the beam smaller and even more precise.

The only thing that remains perplexing is that old conundrum of how our own speed toward or away from a light source fails to change its measured velocity. Why should photons from a rapidly approaching Corvette headlight hit our testing device at the same speed as those from a parked car? It's as if we were to stick an arm out of that zooming race car and feel the air remain dead calm. It doesn't make sense.

As far as light is concerned, we're still back in Genesis on a stationary Earth.

As far as light speed is concerned, the motion of everything else doesn't exist.

Everyone contemplating this, including generations of physicists and science lovers, shakes his or her head in wonder and disbelief.

CHAPTER 16: *Meteor in the Kitchen*

And Other Peculiar End Points

The untented Kosmos my abode,
I pass, a wilful stranger:
My mistress still the open road
And the bright eyes of danger.

—ROBERT LOUIS STEVENSON, "YOUTH AND LOVE: I" (1896)

B arely more than a century ago, on June 30, 1908, a black-bearded man sat on the front steps of his cabin in one of Earth's most isolated regions. It was exactly 7:14 in the morning, though he had no knowledge of the time, because he owned no clock. Tall, straight pines surrounded his modest home in south-central Siberia, built without the benefit of power tools. The spot was chosen because of plentiful water from an adjacent brook in this animal-abundant region northwest of Lake Baikal.

Then and there, he witnessed the largest meteor impact in recorded history.

A brilliant blue ball, nearly rivaling the sun, "split the sky in two." But unlike the bolides, or exploding meteors, that lucky observers may witness every few years over their own homes, this one neither vanished at the horizon nor fizzled into a shower of sparks. Instead it simply exploded in the air forty miles northeast of where he stood gaping, shielding his eyes against the low morning sun that was visible in the same general direction.

Immediately he felt intense heat on that side of his body, "as if

238

my shirt was on fire," he explained years later. He wasn't sure if he should rip off his clothes; wouldn't his bare skin then be perilously exposed to whatever this was? His indecision ended when he heard a loud thump, the earth shook, and "hot wind raced between the houses like from cannons, which left traces on the ground like pathways." This violent wind immediately blew him sideways into the air and hurled him ten feet. He lay in the dirt, barely conscious.

When the first Soviet expeditions reached the ground-zero area more than ten years later (a larger scientific team arrived in 1927, a delay understandable in those tumultuous times), they found eight hundred square miles of utterly destroyed, charred trees that had been blown down radially, all pointing away from a spot three to six miles beneath where the intruder exploded. Later analysis showed that it had been either a small comet or stony asteroid the size of a large house. Its explosion had released a force of between five and fifteen megatons, rivaling a thousand Hiroshima bombs.[1] Just a tiny comet or asteroid fragment. Smaller than a movie theater. Nothing out of the ordinary. Its speed alone had made it dangerous.

The worrisome thing is that it has two hundred thousand cousins.

In what became the strangest celestial coincidence in recorded history, a somewhat smaller air-bursting meteor exploded over Siberia once again, this time on February 15, 2013. We know it was smaller (probably the size of a bus) and blew up when it was higher, because this one hurled no one through the air and didn't knock down a single tree, though its shock wave broke many windows, causing a thousand injuries. It was a coincidence not just because the target was Siberia once again but also because this occurred on the same day as the closest-ever observed flyby of a substantial (football field–size) asteroid—which missed us by a mere seventeen thousand miles.

There had previously been only a single rigorously documented instance of human injury from a meteor. Suddenly, in 2013, the world's total casualty list from extraterrestrial objects went from one person hurt in five thousand years of recorded history to a thousand people.

Nothing in the universe is stationary. Absolutely everything moves.

We don't even have to cast our nets to galactic distances. The velocities we can probably most easily relate to are those in our immediate neighborhood. These entities affect us. The moon and sun align either together or on opposite sides of the heavens—both configurations exert an identical "pull"—to create maximum tides every fortnight. Creatures that depend on the tides, such as clams—especially those in the extensive marine marshes—and their predators, such as gulls, display behavior patterns attuned to the four daily tidal extremes and also to the larger biweekly ones. Celestial rhythms thus echo from the skies and spill over into the animal kingdom.

Remaining relatively near to Earth, to the places our robot surrogates have visited, we might begin with one of the four celestial bodies on which humans have left garbage: the moon.[2] It so happens that the moon is one of the cosmos's slowest entities. The moon requires four weeks to make a single rotation. And, famously, it orbits around us in precisely that same amount of time—27.32166 days. But this seemingly bizarre coincidence turns out to be logical and commonplace. Nearly all the 166 moons in the solar system—which mostly have names like Greip and Puck and Neso, which few people have ever heard of—rotate and revolve in matching periods, too. Their months and days are the same.

It simply means that when two celestial bodies share the same neighborhood, they influence each other. The larger one's gravity dominates the system and exerts a braking tidal action on its neighbor. The smaller one's spin gradually slows until one hemisphere is locked in place, this half forever facing its parent planet. Thus our

The moon is the only object in the known universe that moves its own width in one hour. Here, in a photograph taken in the Libyan desert in 2006, it has just finished using that hour to fully cover up the sun. *(Terry Cuttle)*

situation of always seeing one familiar side of the moon, with markings that seem glued in place and with a hidden hemisphere perennially pointing away, is as common as tuna on white.[3]

This doesn't mean the moon is devoid of cool animation. Far from it. It's the only celestial body whose speed through space is "one diameter per hour."

This manifests itself to our naked eyes and always has. During an eclipse, when the moon either passes into Earth's shadow or blocks the sun, it duly requires very nearly one hour to become immersed. And on any night, our nearest neighbor shifts one "moon width" against the background stars every fifty-seven minutes as it chugs through space at 2,289 miles per hour. (This is its average. It can move 126 miles per hour faster or slower in its oval path around us.)

Motionwise, the nearest planets to Earth — Venus and Mars — play the character roles of George Burns and Gracie Allen. They are Mr. Normal and Ms. Peculiar. Mars, the straight man, has a spin remarkably similar to ours. Its day is 24.5 hours long. But Venus is strange. It is the slowest-rotating object in the universe. A Venusian day is 244 Earth days. It barely spins at all.

Confining our musings to our four nearest planetary neighbors, we discover that Jupiter provides a fantastic contrast, because it's the *quickest*-turning body. Despite its huge size—1,300 Earths could fit inside a hollow Jupiter—it rotates in just under ten hours. This is a planet with attitude. Its equator moves twenty-four times faster than ours. So fast that its clouds are horizontal streaks and resemble paint thrown onto a spinning turntable—especially when viewed by spacecraft from above either of Jupiter's poles. That planet whirls so fast that its equator bulges outward, making Jove not remotely round but squashed at the poles.

To visualize these disparate planetary spins, picture yourself strolling along each equator's bike path. On Venus, simple walking speed would be enough to outpace its rotation. A brisk promenade would keep night from ever falling.

On the moon you'd need to trot a bit faster, but still a marathoner could keep the sun from setting. Its rotation is just ten miles an hour. That's the speed at which the shadows cast by lunar mountains creep along crater floors. But Jupiter's equator zooms along at twenty-five thousand miles an hour. Fifty times faster than a bullet.

In addition to its spin speed, each planet has its own unique forward speed as it orbits the sun counterclockwise (as seen from above, or north of, the solar system; all planets move in the same direction, a one-way procession).

Planet speeds have an easy, logical sequence. The rule is simple. The closer you are to the sun the faster you must move to maintain a stable orbit and not get pulled into its gravitational field and become vaporized. Tiny Mercury zooms along at thirty miles a second.

Venus whizzes at twenty-two miles a second. Our world moves at 18.5 miles per second. Mars goes at fifteen miles per second. You see how it works. Planets farther from the sun move more slowly. Poor demoted Pluto lopes along at just three miles a second.

In our celestial vicinity, none of the speeds particularly stands out. None is off the charts. The planets are like adjacent sprinters on a circular track, each in its own lane. The nearest planet to us on the sun's side goes just 3.5 miles per second faster than we do. The one on the outside goes 3.5 miles per second slower. Nobody moves too crazily compared to its neighbors. That's why meteors streaking across our night sky seem fast but not insanely so. Nearly all of them are broken pieces of comets or asteroids whose parents orbit the sun in our general vicinity. Their speeds resemble ours.

Meteor watching is always fun, especially between midnight and dawn, when six shooting stars reliably tear across the sky every hour. If this isn't enough for you, Earth intercepts a thick swarm of comet debris several times a year, and then it's an all-you-can-eat affair. We then see sixty or more meteors an hour if we observe away from city lights. One a minute. These meteor showers — on August 11, December 13, and sometimes on November 18 — offer vivid motion demos, and here's why:

The speed at which a meteor rips through our atmosphere depends, oddly enough, on its direction. This is very different from our experience on earth, where the crosstown bus doesn't race more furiously than the downtown one. But meteoroids' space speeds are all rather similar to ours.

Notice the new term *meteoroids*. Space rocks always seem to change their names. Here's how it works: when a hurtling rock flies through space, it's called a meteoroid. That's what can and does occasionally strike our satellites and even the International Space Station. A stone in space even a foot across, zooming at sixty thousand miles an hour and not giving off any light, is utterly invisible. But if it enters the atmosphere and burns up, it's then called a meteor. We rarely see the metallic chunk itself, since it's typically the size of a raisin or even an apple seed. Rather, we view the glowing, ionized, heated air that surrounds the invisibly tiny white-hot stone. The phenomenon is commonly called a shooting star. Finally, if the meteor does make it to the ground and gets discovered, it undergoes another name change. Now it's a meteorite.

In any case, a meteoroid's speed is less crucial than its direction. What matters is whether it's barely catching up to Earth from behind or instead hitting us head-on. That's what's all-important. The August 11 meteors, those famous Perseids, smash into us head-on. So we witness the violence of their orbital speed added to our own for a combined impact velocity of thirty-eight miles per second. The November Leonid meteors do the same thing. These screaming shooting stars streak across the sky in a mere second or two. No time to say, "Hey, look at that!" Glance down and you miss them.

But the December 13 Geminid meteors arrive here at a 90-degree angle relative to our orbital direction. There is no head-on collision. They're like a car backing out of a driveway and crunching softly into ours from the side. The impact speed is just half that of the others. They only streak at twenty miles a second, and it shows as they lope lazily across the heavens. In the vast majority of cases, they are even too slow to create glowing trains behind them, unlike fully one-third of the Perseids and Leonids. What's wild and satisfying is how easy it is, with no telescope or any other equipment, to witness these disparate cosmic velocities.

That 1908 Tunguska meteor was moving from the east toward the north. At that early hour of the morning, this corresponded to a sideways entry into our atmosphere. Similarly, the Siberian bolide (exploding meteor) of 2013 came in at about eleven miles a second from the sunward direction, sideways to us. Had either come from an overhead direction it would have had far more speed and thus released far more energy. Our thin, dirt-covered 1908 witness, Semen Semenov, whose wife led him back into his house, the windows of which were now broken, may not have fared as well if that had been the case.

Obviously, celestial movement isn't necessarily a textbook sort of thing. It parades overhead right in our faces when we take a few minutes to look.

But if celestial speeds do seem too cerebral, we can bring the whole thing home, quite literally. Remember that our planet is not an isolated island. Comets and asteroids continue to pay us close encounters of the clobbering kind.

The public has many misconceptions about *meteorites*—the name for meteors that land. People imagine they're hot when in truth they're barely warm, after being flash-frozen by passage through our cold atmosphere. On August 31, 1991, two boys standing on a front lawn in Noblesville, Indiana, saw a meteorite thud into the grass a few feet away and picked it up immediately without harm.

People also imagine they're deadly, when the only person ever directly hit, Ann Hodges of Sylacauga, Alabama, was merely bruised when one tore through her roof, struck a console radio, then bounced onto her hip on November 30, 1954.

To merely state that Ann Hodges was the only person in history to be hit and injured by a meteorite is to simplify an amazing story. It started that afternoon when Ann, not feeling well, fell asleep on her living room sofa—in a rented white house located across the street from the Comet Drive-In Theatre, whose neon logo depicted a zooming, meteorlike object.

Hodges was awakened by the eight-pound metallic object crashing at high speed through the living room ceiling. Before she could react, it bounced off the radio, struck her left hip, and bruised her left hand. The incident quickly drew crowds of TV and print reporters and put the thirty-four-year-old woman in the history books. A footnote was added for area physician Moody Jacobs, the only doctor ever to treat someone struck by an object from outer space.

But Ann Hodges did not gain any benefit from this historic occurrence—unlike, say, eighteen-year-old Michelle Knapp of Peekskill, New York, whose life was changed by a 1992 meteorite encounter with her car. For Ann, the trouble began when her husband, Eugene Hulitt Hodges, and she were upset at the crowds descending on their home, then amazed and angered that police officers and government officials took away the meteorite without the family's permission.

The Hodgeses worked with a lawyer to eventually secure the meteorite's return, but their hopes of making a fortune from the stone quickly faded when their landlady, Birdie Guy, claimed the meteorite was rightfully hers and fought for its custody in court. Legal battles over its ownership and multiple costly appeals all went against the Hodgeses, while public sentiment went against the "greedy" landlady, as she was generally depicted in news reports. The Hodgeses finally settled, and Guy accepted five hundred dollars in lieu of the meteorite, but by then the headlines were long over and the meteorite was no longer a hot or valuable item. The couple eventually turned it over, for small compensation, to the Alabama Museum of Natural History at the University of Alabama in Tuscaloosa, where it remains on display.

The only person with a positive experience in the entire celestial encounter was a farmer named Julius Kempis McKinney. On December 1, 1954, the day after the meteorite struck the Hodges house, McKinney was driving a mule-drawn, firewood-laden wagon a few miles away when the animals balked at a black rock in the road. McKinney kicked the odd black stone off the road and continued home. But later, upon hearing news reports of the Hodges incident, McKinney returned to the site. He took the rock home and let his children play with it.

He trusted only his postman with the information. The postman helped McKinney find a lawyer, who negotiated an amazing price for the meteorite's sale. The purchaser was an attorney from Indianapolis acting on behalf of the Smithsonian Institution.

Mineralogy experts confirmed that the three-pound rock was indeed a fragment of the larger Hodges meteorite; it is actually common for a meteoroid to shatter or explode into multiple pieces in the air before hitting the ground. Although the sale price was never revealed, it was enough for the McKinney family to purchase a car and a new house. This unexpected good fortune was a rare event for an African American in that state in that period of history, when racial inequality was the norm.

Is all this sufficient for a movie plot? Bill Field, who as a five-year-old in Sylacauga saw the meteor streak across the sky, leaving a white trail, and heard a loud sonic boom—a somewhat quieter version of what Siberian townspeople experienced in 2013—grew up to be a filmmaker. He researched the incident and what happened to all the people involved and successfully sold his movie script to 20th Century Fox. But no film was ever made.

As for Ann Hodges, she later said she'd been permanently changed—not by the six-inch bruise on her left hip but by the emotional scars resulting from the legal fights and disappointments. She died of kidney failure at a Sylacauga nursing home in 1972, at the age of fifty-two.

History also records a Franciscan friar who was reportedly killed in Milan in the seventeenth century by a two-inch meteor that severed an artery in his leg. But who can say for sure? For all anyone knows, it could have been a stray musket shot. Confirmation is lacking. And in 2009, a German boy claimed that his finger had been injured by a pea-size meteorite that appeared "after a flash of light" and then "buried itself in the road." Despite global headlines, however, this story is not credible at all. Ann Hodges's bruise remains the only authenticated human injury from a hurtling space object.

Meteors provide the only visible physical interchange between Earth and the heavens. It's the sole visual link between what's "up there" and our terrestrial lives. A meteor's fall to earth is sudden, and just to add a little spice, there's even that hint of danger.

Hint or not, Armageddon lovers never tire of the "peril from space" motif—the fear of a giant Earth-smiting stone from hell, a Tunguska meteor on steroids. New end-of-days predictions sprout like poppies, only to dissipate when the dreaded date passes harmlessly and is as promptly forgotten as an opium dream. To appreciate the peril realistically, you have to know how it works.

Instead of watching meteors in space—few of which survive their trip through the atmosphere but instead disintegrate quietly into dust—we should contemplate the rare dramatic meteors that have made it all the way to the ground. We won't even talk about the really bad events that altered history, including the dinosaur-ending K-T impact sixty-five million years ago, which crashed into the raptors' favorite Yucatán beach at Chicxulub. Or the even worse Permian "great dying" 251 million years ago, which destroyed most of the planet's genera as though they were doodles on a chalkboard and nearly erased the intricate, interconnected biosphere. Such events typically involve asteroids more than a mile wide, the kind that seem to hit us every few hundred million years or so. Much more common are the smaller stones in the three- to twenty-five-pound category, which damage homes almost yearly, usually after an expensive renovation.

Meteoroids that make it to the ground are not the meteor-shower variety, which are usually composed of flimsy ices. The survivors are, instead, hardy stony or metallic pieces of asteroids or even fragments of the moon or Mars. They arrive without warning.

A meteoroid can weigh a ton as it strikes our atmosphere; that

248

was the estimated mass of the intruder that broke into dozens of fragments over a Chicago suburb on March 26, 2003. One small piece penetrated a teenager's bedroom, hit his printer, and shattered a full-length mirror. Bad luck? It could have been much worse.

A meteoroid traveling through space encounters Earth at a sizzling seven to forty-four miles per second. If it weighs more than one hundred thousand tons our atmosphere won't slow it down in the slightest: it slams into the ground at full cosmic velocity.

At the other extreme, if the meteoroid is less than eight tons, friction from the air robs it of *all* its original speed. Then its impact, like that of falling garbage or flying squirrels, is strictly determined by terminal velocity. Happily, these lower-mass objects are the rule.

At a height of about ten miles, or fifty thousand feet, a meteorite slows to two or three miles per second and no longer glows. From that altitude on down it's an invisibly dark tumbling piece of rock, which can be mostly stone or a half-iron, half-nickel amalgam. Nonetheless its seven-thousand-mile-an-hour velocity, three to six times faster than a bullet, gives a one-pound meteor enough kinetic energy to bring down a jetliner. It hasn't yet happened, but it could.

Continuing downward, unobservable, the meteoroid's encounter with increasingly thick air slows it to a terminal velocity of around 250 miles per hour. This is its final speed as it strikes the ground. Or anything else.

Buildings are penetrated every year or so in North America alone. Animals, standing naked outdoors, have also fared poorly:

May 1, 1860: A horse is killed by a meteorite in Concord, Ohio.

March 11, 1897: A rain of many stones in West Virginia kills another horse.

June 28, 1911: A meteor later found to have come from Mars kills a dog on the outskirts of Alexandria, Egypt.

October 15, 1972: A cow is killed by a meteorite in Valera, Venezuela.

Cars seem to attract meteors, too. One was quietly parked in its garage on September 28, 1930, in Benld, Illinois, when a meteor penetrated the garage roof, the car's roof, and the car's wooden floorboards before bouncing up off the muffler and coming to rest in the seat material, inaugurating a long love affair between asteroid fragments and automobiles.

In the past quarter century the most spectacular encounter was the parked Chevy in Peekskill, New York, whose trunk was destroyed by a twenty-six-pounder on October 9, 1992. Its eighteen-year-old owner found her life changed by the sixty-nine thousand dollars paid to her by a collector. (He wanted the crumpled car as well as the meteorite, and she said, "Sure." A beat-up, ten-year-old Malibu whose insurance didn't even cover the damage? Are you kidding? Take it!)

Just between 2002 and 2010, meteorites entered at least seven homes, including two in the United States. The freshly landed stones are usually "barely warm" and have a black fusion crust.

There is no terminal velocity on airless bodies such as Mercury and the moon, where meteoroids captured by gravity keep gaining speed up to a maximum that equals that planet's escape velocity. On Earth it's twenty-five thousand miles per hour. If there were no air, meteorites wouldn't just penetrate roofs and floors. They'd keep going until they converted the basement playroom and much of the surrounding neighborhood into a huge crater. After all, kinetic energy equals the meteorite's mass times the *square* of its speed. A meteor reaching your kitchen table at 250 miles an hour is 100^2, or *ten thousand,* times less damaging than if it had struck at even a low space speed of twenty-five thousand miles per hour.

That's why meteor stories have, to date, tended to be whimsical (or, in the case of the 2013 Siberian event, scary and laceration-producing) rather than tragic.

Meteor landings, such as the dozens that occurred in a Ugandan village in 1992, are sometimes preceded by widely observed celestial fireworks. My favorite meteor story involved just such a scenario in the northeastern United States on November 30, 1981.

An alarmed woman phoned our observatory (the Overlook Observatory, which I have owned and operated since 1982) that night to report a fiery ball slashing across the heavens, lighting up the countryside. Some people assume that observatories are UFO-reporting stations, and we get regular inquiries concerning lights in the sky. But like most celestial phenomena, this had an easy explanation, and I told the woman that the sparking object was probably just a meteor: nothing unusual. I couldn't know, however, that things were anything but routine a mere hundred miles to our east.

Observers in central Connecticut were noticing the same brilliant light in the sky, but to them it was motionless. There's only one way it could appear stationary: it was coming straight toward them!

The grapefruit-size meteor not only survived its passage through the atmosphere, it crashed through the roof of a house in Wethersfield, Connecticut, where Robert and Wanda Donohue were watching the TV show M*A*S*H* in the next room. They later told me that it was the loudest sound they'd ever heard. Rushing into a room that was now filled with dust, where furniture had been overturned, they found a hole in the ceiling.

Connecticut has no Meteor Police, and after the Donohues called 911 some firemen came to their house along with the town's uniformed officers. It was a fireman who found the six-pound meteor under the dining room table, where it had settled after a couple of high-speed bounces that left behind scuff marks on the carpet and ceiling.

They almost weren't surprised. Eleven years earlier, in April of 1971, the last time a meteorite had hit a house anywhere in the United States, the impact point was Wethersfield, Connecticut. The same town. In one of the strangest coincidences of our time, a house barely more than a mile from the Donohues' had been struck.

The only plausible explanation for the same town being hit consecutively is that Wethersfield is a suburb of Hartford, the headquarters for many insurance companies. This is where statisticians and actuaries live. They're the ones who know how impossible this is.

(In case you're curious, the answer is yes: the Donohues' insurance completely covered them for the meteorite damage. They deserved it. A couple of years later, Bob and Wanda generously donated the cosmic house wrecker to a museum in New Haven.)

With all these ongoing impacts, should we worry, really? Maybe a little. That famous Tunguska event occurred over a part of Asia at a time when the world's population was just a third of what it is today. If it happened now, over a city, we could have twenty million fatalities.

Significant meteoroids keep coming in, like the six-foot-wide exploding air burster that rattled windows in Nevada on April 22, 2012, and of course the 2013 Siberian spectacle. Gemologists and adventurers quickly converged on both places like Black Friday shoppers and started finding meteorites—most of which bored precise holes into the snow—in Northern California the next week and in and around the Russian town of Chelyabinsk the very next day. But experts now estimate that a truly damaging meteorite impact hits our world only every few hundred years. Then and now, the most likely ground zero is somewhere over the ocean.

The asteroid Apophis will come extremely close to us on April 13, 2029, at a speed of nineteen miles a second. It will barely miss us then, passing between the ground and our television satellites 22,300 miles up! If its orbit is altered in a precise but unlikely way by that near miss, it could hit us the next time it comes by, on April 13, 2036, with an impact explosion equivalent of five hundred hydrogen bombs. However, the chance of that collision with Earth is currently pegged by NASA experts at only one in a quarter million, which matches the odds of your teenager grabbing the vacuum and spontaneously cleaning the entire house.

Far beyond our solar system, truly off-the-charts velocities— like those of galaxies slamming into each other at a few percent of the speed of light—will never affect us. The cosmos is crammed with rapid motion for the mind's musings only, an insurance headache for alien civilizations alone.

The fastest you and I—and everything else on our forgiving planet—travel through space? Aristarchus nailed the twin spin-plus-orbit motions 2,300 years ago. Combined with Eratosthenes's spot-on determination of our planet's size a century later, the small minority of humans who eschewed geocentrism knew before Christ was born that Earth spins.

When the four eighteenth- and nineteenth-century transits of Venus across the solar disk let astronomers pin down the sun's true distance from us, we could then finally calculate our exact orbital speed: 66,600 miles per hour, or 18.5 miles per second. We'll never feel it, because everything around us is moving, too, plus there's no palpable acceleration or motion change.

Only one other major earthly speed needed to be added. The all-time biggie. This was uncovered by Harlow Shapley a century ago. As we circle the sun, that star itself whooshes around the galaxy's own center, taking us along for the ride. Our world thus partakes of our galaxy's spin at a whopping 140 miles a second. That's the very fastest terrestrial speed that makes any sense, because beyond the galaxy there's no fixed reference point. We say the

Andromeda galaxy is approaching us at seventy miles a second. But we could just as easily regard it as motionless and say that we are moving toward it at that speed. Or we could split the difference and say each travels at thirty-five miles per second. All we know is that the gap between us is shrinking. Because we lack any stationary reference grid for extragalactic motion, our ability to include calculations of speed in our movement story stops at the property lines of the Milky Way. Beyond that, spaces between galaxy clusters increase, but no one can pin down who, exactly, is moving.

Old textbooks say that the sun and Earth move together through space at just thirteen miles a second. That's because, not too long ago, we were only aware of our motion *relative to the stars surrounding us*. Imagine a group of floating leaves rushing down a river's rapids. One leaf has a bit of a slow, sideways drift relative to the others. That's what those old books talked about. Like adjacent horses on a carousel, the stars in the night sky—which on average are just 150 light-years away—partake of the same motion we do. So they don't seem to move much, relative to us. Observing them, we seem to be slowly drifting at thirteen miles per second toward the star Vega (some authorities peg it as the constellation Hercules in that same part of the sky). However, we know now that we, Hercules, and Vega all simultaneously rush crazily forward at 140 miles a second in the direction of the star Deneb, which we'll never reach, because it's moving ahead at the same rate.

Although this ultimate sensible motion of our world unfolds ten times faster than our best rockets, it's still just one-thousandth the speed of light.

CHAPTER 17: *Infinite Speed*

When Light's Velocity Just Won't Get You There

*I could be bounded in a nutshell, and count
myself a king of infinite space,
were it not that I have bad dreams.*

—WILLIAM SHAKESPEARE, *HAMLET* (CA. 1600)

There's fast, and then there's infinite.

Plenty of things are fast. All the atoms around us vibrate trillions of times a second. Photons in fiber-optic cables completely circle the earth in a literal eyeblink. Distant galaxies whoosh 150,000 miles farther away each second.

Infinitely fast is a different ball of wax. It would mean that something leaving the farthest galaxy just as you reach *this* point of the sentence is now already in Kansas. We always thought such superluminal speeds were impossible. We were wrong.

Infinity's exploration requires a quick peek into the intriguing realm that surrounds light speed, which seemed like the absolute limit when many of us went to school.

In 1905, Einstein explained a wild observation made two decades earlier by Hendrik Lorentz and George FitzGerald. They had realized that light travels at a constant speed and understood how profoundly remarkable this is.[1]

It means photons from the landing lights of an approaching jet strike you at light's unwavering rate of 186,282.4 miles per second, as if the plane weren't moving at all. Right from the get-go, light starts out as unique and nonintuitive.

Moreover, Einstein showed that objects that have weight can never quite reach light's speed. In a hypothetical ultrapowerful rocket, as you accelerate your mass increases. You magically get heavier and heavier. At just below the speed of light, even an object that started out lighter than a feather would outweigh the entire universe. The energy needed to accelerate it the final tiny amount would be infinite. Hence you could never achieve that speed.

After Einstein set forth his two relativity theories in 1905 and 1915, light's sovereignty in a vacuum was no longer seriously challenged. Yet bizarre escape clauses started to show up during the ascension of quantum mechanics in the 1920s.

This is a Wonderland realm where objects don't quite exist until they're observed. There are two main competing theories attempting to make logical sense of this. The first is the "many worlds" explanation for quantum phenomena. This maintains that each option in life creates a separate universe that then carries on. The moment an alternative possible action exists for anything—even if you observe a falling leaf landing here but not an inch away—the cosmos divides into separate realities to accommodate both outcomes.

If you measure an electron, you've deliberately or unintentionally forced it to appear in a particular place with particular properties, such as an upward spin versus a downward spin. Or, to be more accurate, you have suddenly joined the universe where it exists in the state you observe it to be. But different yous also exist, inhabiting separate universes, where they each observe the electron in all the other places or states that were then possible.

By this reasoning, some other version of you really did take the prettiest cheerleader to the prom. Unfortunately, one analogue of yourself was a jerk that night (remember, if it *could* happen, it *did* happen), and she never spoke to that version of you again.

Most theorists and science professionals do not buy into all these simultaneous realities. Instead, the majority prefers the *Copenhagen interpretation.* This does away with multiple realities but says that the universe is filled with particles and bits of light that have no definite existence, location, or motion until they are observed. Only then does their wave function collapse, and only then do they materialize in a statistically determined place and continue to exist there happily from that moment on.

Einstein didn't like any of this. In 1935, he and two colleagues, Boris Podolsky and Nathan Rosen, wrote a now-famous paper in which they essentially bad-mouthed quantum theory as fundamentally incomplete and thus seriously flawed and addressed an aspect of quantum theory that was bizarre even by quantum standards. They pondered what happens to particles created together, or "entangled." According to quantum thinking, the pair of particles then *shares* a wave function, and each object knows what the other is doing. If one is observed, forcing it to leave its blurry, probabilistic wave-function state and collapse into an electron with an "up"-pointing spin, its twin—no matter where in the universe it happens to be—knows what its doppelgänger did, which causes its own wave function to collapse. It instantly becomes a particle with complementary properties, in this case a "down" spin.

The easy way to create such entangled pairs is to shoot a laser into beta barium borate or certain other crystals. Suddenly *two* photons emerge, each with half the energy (twice the wavelength) of the original, so there is no net energy gain or loss. These two then head off at the speed of light, possibly for billions of years, to lead seemingly independent lives. The same process holds true for entangled solid objects such as electrons and even whole atoms and clumps of material.

But let one member of the duo collapse into a particular state, and its twin knows this is happening and instantly does the same.

Einstein, Podolsky, and Rosen argued that such apparent

parallel behavior must be attributable to local effects, a contamination of the experiment, rather than some sort of "spooky action at a distance," as they called that aspect of quantum theory. Their paper was so celebrated that such synchronized quantum antics borrowed the physicists' initials and became known as EPR correlations. And the line "spooky action at a distance" became the standard pejorative way of describing such an outrageous and silly belief—a putdown of true instantaneous behavior. It was repeated in dismissive fashion in physics classrooms for decades.

But recent experiments show that Einstein was wrong. In 1997, Geneva researcher Nicolas Gisin created pairs of entangled photons and sent them flying apart along optical fibers. When one encountered the researcher's mirrors and was forced to go one way or another, its entangled twin, seven miles away, always instantaneously acted in unison and took the opposite, complementary option when encountering its own mirror.

Instantaneous is the key word. The reaction of the twin was not delayed by the amount of time light could have traversed those seven miles to convey the news. It happened at least ten thousand times faster, which was the experiment's testing limit. Quantum mechanics tells us that the echoed behavior should indeed be perfectly simultaneous. Indeed, quantum theory predicts that an entangled particle knows what its twin is doing and *instantly* mimics its actions, even if the pair lives in separate galaxies billions of light-years apart.

This is so bizarre, with implications so enormous, it drove some physicists to a frantic search for loopholes. Some argued that Gisin's testing apparatus had a bias and preferentially was detecting only those particles that exhibited the complementary properties expected of twins. Then in 2001, National Institute of Standards and Technology researcher David Wineland eliminated these criticisms.

Wineland used beryllium ions and a very high detector

efficiency to observe a large enough number of events to seal the case. So this fantastic behavior is a fact. It's real. But how can a material object instantly dictate how another must act or exist when the two are separated by large distances? Few physicists think that some previously unimagined interaction or force is responsible. Striving to understand, I asked Wineland what he believed, and he expressed an increasingly accepted conclusion.

"There really *is* some sort of spooky action at a distance."

Of course, we both knew that this clarifies nothing.

So particles and photons—matter and energy—apparently transmit knowledge across the entire universe instantly. Light's travel time is no longer the limit.

Some physicists say that this does not violate relativity because *we* cannot exploit this to send information faster than light, since the "sending" particle's behavior is governed by chance and not controllable. Moreover, nothing of any mass is making the journey. Indeed, nothing weightless, even a photon, is making that infinite-speed journey, either. But *something* is being conveyed instantaneously.

The scientific (not to mention philosophical and metaphysical) implications are astounding. Let's say some of the atoms in your body originally formed in an entangled manner with other particles soon after the big bang. Since then, both have been flying apart, and now they are separated by billions of light-years. Your atoms make up pieces of your brain, which is physically located in Peoria. Those other particles have become part of an alien on a planet in the fashionable Aldebaran system.

Right now, some creature there is observing your twin's atoms in a lab. Bingo, they collapse to exhibit specific properties. Instantly, with no delay whatsoever, your own brain's atoms know this is happening five billion light-years away, and they, too, collapse into complementary objects. The effect is sudden and alters your thought processes, and you make a snap decision. You show up at your boss's party wearing an embarrassing polka-dot tuxedo. You can't explain why you acted so oddly, but your life is ruined. This seems like science fiction, but EPR correlations are real.

First it means that the entire universe is a single entity in some fundamental way. It means there are no secrets between locations here and those far away, no matter how distant—and that the information "exchange" happens simultaneously, at infinite speed.

It means that Einstein was dead wrong about *locality*.

Locality is important in any exploration of motion. After all, movement always implies that things are pushed, propelled, or jostled by other objects or forces, such as wind, water, and gravity. This is what Einstein believed—that an object is influenced only by its immediate surroundings. It's called the principle of locality.

A kind of supplementary principle is *local realism*. This means that all objects have actual properties independent of any measurement of them. An atom, or the moon, is really "there" in some location, and with a definite motion, regardless of whether people are observing it. It's our job, if we're so inclined, to set up ways to learn about this object and to measure its properties.

Contrast this with quantum theory, which denies locality. It insists that an atom can be influenced by events (such as its entangled twin's wave function collapsing) that are not only utterly out of contact with it but on a different side of the universe. And that such influences occur instantaneously. No "carrier particle" is necessary to bring the news or effect the influence from one place to another, nor is the influence limited to some speed, even the speed of light. Instead it jumps in less than an eyeblink from distant empires.

As for local realism, quantum theory does away with that, too. Its popular Copenhagen interpretation insists that the entire universe is made of countless particles such as electrons that *have no inherent location*. Nor do they have any motion. In a real sense, they do not even enjoy any form of true existence. They instead dwell in a kind of blurry probability state of *potentiality*, with tendencies that are statistically decipherable. Upon observation, they materialize according to probability laws.

Einstein indeed hated this. It meant that nothing existed or moved unless observed. It also meant that no one could pin down the actual behavior of individual objects—we could only speak about them statistically as a group and assess the *likelihood* of them being here but not there or moving this way rather than that. This is what caused him to utter his famous antiquantum line, "God does not play dice."

If we set up an apparatus that allows us to detect a particle's location, the object obligingly materializes in a particular place. Yet it still doesn't have a specific motion. But if we instead construct a device that can detect motion, we duly observe the entity to be moving, and yet its position at any given moment is blurry and poorly defined. We can't precisely see its location *and* its motion.

At first scientists thought that this must be a result of some technological immaturity on our part—that if our equipment got better, we'd be able to pin down the motion *and* the location, the way we can with large bodies, such as Saturn. Eventually we came to see that the problem lies much deeper. The small entities that make up everything in the cosmos do not each *have* a location or a movement. Moreover, only our act of observation brings one or the other into existence.

The reason large macroscopic objects do appear to dwell in specific places *and* have motion is because they're composed of so many countless small objects that the overwhelming probabilities of each yield a statistically certain collection in the spot we're observing.

That statistical business is wild, too. While objects normally appear in the most likely places, there's always a tiny statistical chance they'll behave oddly. That they'll materialize far from where they're expected.

Consider a newly paved road with a fresh temporary covering of gravel. Passing cars cause each stone to jump into the air and land somewhere. There's a fifty-fifty chance that a rock being popped up by a tire can go toward the road's edge as opposed to bouncing toward the center. The ones that happen to go toward the edge now have a fifty-fifty chance of flying even farther toward the edge when they're popped up by the next car. Over time, all these probabilities play out, and the road is totally clear of gravel. It's *all* now entirely off the edge—because once a rock is removed from the road the game ends for that rock, which doesn't move anymore. When enough tires have passed, even the pebbles that defied the odds and kept improbably bouncing back toward the center have finally yielded to a series of edge-oriented jumps. The proof is right there: a mere two weeks after the road opened to traffic, no gravel remains. Given enough time, all statistically possible events come to pass, even unlikely ones.

But look closely. Here is one rock that somehow caught the edge of a truck tire, was scraped by an adjacent stone in a very unlikely way, and was propelled hundreds of feet into someone's bird feeder. This single act would probably not have been predicted. It was *extremely* unlikely. But it was *possible*. And given enough time, if there are enough objects involved, all possibilities, no matter how remote, come to pass.

In quantum theory's Copenhagen interpretation, a milk container in your fridge contains particles whose locations are blurry and probabilistic. It's made of many more atoms than the number of gravel stones on a road. (A one-gallon milk container contains the same number of atoms as there are lungfuls of air in Earth's atmosphere.) When you next open the fridge, it is extremely likely that all the container's atoms will be present and that the carton will be sitting where you placed it the night before. Even if one atom materializes somewhere else it would not affect the container's existence on the same shelf as the one where you remember placing it. But it is possible, not impossible, that *all* the atoms will materialize in a most statistically unlikely location. If so, the container will be gone. Perhaps it will suddenly appear in a bedroom in Myanmar.

The chance of all those particles acting in unison in so statistically improbable a way is so small that it is unlikely to happen even in the five-billion-year bionic lifetime of the planet — the period from first bacteria to eventual sterilization. But the point is: it *could* happen. If it does, we see an apparent miracle. We have then observed *motion without any apparent cause.*

So this crazy stuff is true. The observer and the universe go together. And occasional impossible motion is not impossible after all.

Since quantum mechanical behavior and most of the motions discussed in this book involve random activity, it may be worth taking a moment to examine the power and limits of randomness. The usual clichéd example is the monkeys-and-typewriters thing. You've probably heard it: a million monkeys typing for a million years would eventually create the works of Shakespeare just by random chance. Is it true?

In 2003 a research team at a university in England placed a bunch of typewriters in front of a group of six macaques in a zoo enclosure for a month to see what would happen. The animals typed virtually nothing. Instead they pushed food and dirt into the keys, threw some of the machines on the ground, used them as toilets, and quickly rendered all the devices useless. They didn't create any written wisdom whatsoever.

But random actions and probability theory remain a big part of the public's "take" on natural motion. "Chance" is a key aspect of movement to which Aristotle and others gave careful attention. It supposedly has vast powers once it functions freely for long time periods.

So, seriously, *could* a million diligent, dedicated monkeys sitting at a million keyboards for a million years truly create the great works of literature, as is claimed? Believe it or not, such a problem is entirely solvable. Now, keyboards offer a lot of places to push; there are fifty-eight keys, even on old-fashioned typewriters. And 105 or so keys on most modern keyboards. When talking about random events, consider the difficulty of creating merely the fifteen opening letters and spaces of *Moby-Dick:* "Call me Ishmael." How many random tries would be needed?

Given fifty-eight possible keys, the number of attempts would have to be 58×58 fifteen times over, which is three trillion trillion, before success could be expected. With a million never-sleeping monkeys working, all faultlessly typing sixty words a minute (so that typing fifteen keystrokes takes just four seconds), one of them would indeed eventually type "Call me Ishmael."

But odds are it would take thirty-eight trillion years. Three thousand times the age of the universe.

So a million monkeys typing furiously would never even reproduce one book's single short opening sentence. Bottom line: randomness has far less power to achieve results than is popularly imagined.

One other ultrafast superluminal phenomenon may also exist. This one is independent of Copenhagen. In theory, if and when the big bang created the observable universe, it could also have created a cosmos of tachyons—faster-than-light particles. At least it's allowable by math and physics. That's because, although nothing with any mass can ever reach light speed, there is a major escape clause. Namely, the speed limit only applies to objects being accelerated— objects that *start off* slower than light. For them, attaining 186,282 miles a second is a hopeless quest.

But what if, at the universe's birth, there was a realm of objects that went faster than light from the get-go? These tachyons— whose name was coined only in 1967—are permitted by science.

For them, the light-speed barrier remains, but they're trapped on the other side. They can never slow down to light speed!

It takes just as much energy to slow down as to speed up. So tachyons would presumably grow heavier and have their time increasingly distorted as they tried to *decelerate* to light speed.

Like us, they could never cross that barrier. We could never see each other, since photons would never travel from them to us or vice versa. Thus any search for tachyons is a hunt for the invisible.

All this is mentioned only because a study of motion should include a consideration of the fastest conceivable objects. In theory, we should be able to detect the effects of tachyons; they ought to influence cosmic ray showers (also called air showers) and emit detectable blue Cerenkov radiation when they lose energy. Such studies have always come up empty. Few if any physicists believe they exist, even if they remain a sci-fi staple.

So we can probably cross tachyons off the list of things that move. It seems there will be no way to break the photon barrier. Faster than light is out.

Only infinity gets the nod.

CHAPTER 18: *Sleepy Village in an Exploding Universe*

Back Where It All Began

Up a lazy river, how happy we will be
Up a lazy river with me.
—HOAGY CARMICHAEL AND SIDNEY ARODIN, "LAZY RIVER" (1930)

The odyssey was over, and I was back in my repaired home. My never-changing village of two hundred people had not stirred while I was gone. Even the annoying, omnipresent, garden-destroying deer seemed the same, albeit with a few new fawns to carry on the tradition. China may be speedily changing, but you'd have to look long and hard to see anything very different in rural upstate New York during the past forty years I've lived here. Brenda at the post office smiled as she handed me a huge pile of mail, bundled with rubber bands.

My desk and its vicinity were littered with their own explosions of spiral-bound notebooks and loose papers and scribbled interviews. It was wrap-up time. I grabbed the phone to cross-examine the Carnegie Observatories astrophysicist who had promised me results from my night on that mountain in Chile so long ago.

He kept his word. Dan Kelson—still upbeat if a bit disheveled, thanks to his two small kids running around and around his office—was effusive with excitement. The four thousand galaxies he personally measured that night, along with his follow-up observations, had revealed places where cities of stars fly away from us at an impressive percentage of light speed. The measurements had carried him to the strange barrier beyond which humans can never see: the edges of the observable universe.

We were both excited about even newer data. In 2013, he had made world headlines by finding the fastest and farthest galaxy ever. And I had just spoken with Shirley Ho, part of the team from Lawrence Berkeley National Laboratory that in 2012 had finished measuring data from an astonishing nine hundred thousand galaxies. They'd used sound waves propagating through the younger, denser cosmos—called baryon acoustic oscillations—to gain groundbreaking information, which these days pours in at a rate previously seen only in science fiction movies.

"They show, without a doubt," she'd said to me, "that space has a flat topology."

Dan Kelson and I now discussed this with excitement. You see, if the entire expanding universe is finite, with a specific limited inventory of stars and galaxies and energy, it would warp space itself. Light traveling long distances would gradually curve. But this new data supports Carnegie's previous findings: light *doesn't* curve. It travels in laser-straight lines. On the largest scales, space has a flat topology.

This strongly suggests an infinite universe. That there's no end to the galaxies. And—back to motion—that speeds just keep getting faster and faster the farther you look, without any terminal point.

We already observe galaxies that are essentially flying away at the speed of light. And of course we're observing them as they were in the distant past, nearly thirteen billion years ago, when their light started out on their long journey to our eyes. Projecting where they must be today, we conclude that they are *currently* zooming away far *faster* than light speed.

And it just keeps going. How can anyone comprehend this?[1]

* * *

I asked the president of the American Astronomical Society, Debra Elmegreen.

"Yes, we may indeed have a flat topology and an infinite universe," she acknowledged, echoing what Shirley Ho had said a week earlier. "But even if we can only observe a tiny fraction of the whole thing, that still amounts to two hundred billion galaxies. It's quite enough to keep us busy."

No doubt. But she slightly misspoke. "Infinite" space doesn't mean "very large." It doesn't mean that everything we observe is "a tiny fraction" of the actual universe. Any percentage of infinity is zero. So all we can ever observe is *zero percent* of the cosmos.

I felt like Alice, endlessly tumbling. Can it be that the entire tapestry we observe is not even a few brushstrokes of the entire cosmic masterpiece? I contacted Caltech theoretical physicist Sean Carroll, who cautiously said that while our observations would be able to prove a finite universe if it existed, you can never *prove* infinity. Nonetheless, given the current data, he believes that "the universe probably is infinite."

What does that mean regarding cosmic motion and everything else? Well, he said, "either an infinite number of different things happen or a finite number of things happen an infinite number of times. Either of those possibilities is pretty mind-boggling."

An infinite universe—the increasingly likely reality—also means that the most energetic motion event of which we're aware, the big bang, was probably just a local happening, a big to-do in the 'hood, confined to the *observable* universe. As for the larger universe beyond, no one can do more than speculate. Does it simply exist forever? Did it "start" smaller and, thanks to the mysterious dark energy that continuously inflates its expansion rate, ultimately grow into what will *become* infinite size? If one had to bet the farm on it, smart money would wager that the cosmos never even had a birth. Which means Aristotle was right: we're part of an eternal entity.

I phoned University of Chicago cosmologist Rocky Kolb to get his take on all this. He just chuckled. An infinite universe, he said, would have "started out everywhere at once, *as infinite from the beginning.*"

He confirmed that, given the likelihood of infinity, the speed of receding stars and galaxies is *limitless.* For simplicity's sake let's call everything we can observe, to a distance of thirteen billion light-years, "one universe," or 1 u. Infinite expansion, in which speeds increase exponentially with distance, as we're already observing, means that at some faraway location galaxies currently increase their separation from us by one universe per second. Call it 1 ups.

We created our original units of measurement based on human experience. A foot was very nearly the length of a man's shoe, a yard was a long step. A mile was how far a person walks in twenty minutes. Even by these clumsy standards, we are able to state — and perhaps even grasp — that the most distant observed galaxies are something like 170,000 miles farther away from us each second.

But unseen multitudes growing *one universe* farther away per second? And, say astrophysicists, who grasp that the visible cosmos is barely the iceberg's tip of all that exists, this still isn't the end. There must be far more galaxies zooming at the speed of one million ups. One million universes per second. And that's not the end, either.

We've seen the lower terminus of speed. It's absolute zero, where even atoms stop moving except for some subtle quantum effects. You can't go any slower than stopped. But the upper limit, long thought to be light speed, has now been penetrated with a vengeance. The gap between ourselves and those unobservable distant galaxies grows in ways we can never visualize because nobody can picture infinite speed. Exploring cosmology is starting to resemble the medieval study of magic. There's no answer to it.

Is that the last word? Ever-increasing accelerations of limitless objects? All of which are forever unobservable? Fathomless oceans of mysterious entities whose mere contemplation is an encyclopedia of futility? What do we do with this? Should we feel titillated or suicidal?

Happily, there's a catch. Physics shows us that space itself may be real on some levels but not others. Maybe there's something fishy about all this distance and velocity in ways our science has yet to fully grasp. Looking at millennia of cataclysmic changes, of bedrock certainties overthrown even in our lifetimes, we see that virtually everything we now know about movement through the cosmos seems alterable.[2]

"All scientific theories are models of nature based on observation," explained my friend Tarun Biswas, a relativist and physics professor at the State University of New York. "The problem with cosmology is that its current model is based on negligible observational data. It would still not be a problem if people did not take it so seriously—if they understood that it is only a starter model." If we can remember that we are still taking our first baby steps in visualizing the cosmos and its contents and its motions, we may be less frustrated by the mad superluminal recession at its fringes.

The fastest speeds, remember, are not those of material objects being accelerated but that of the empty space expanding between us and them. The modern physics chronicle has a certain disconcerting quality: in the quantum phenomenon of tunneling, objects pass through supposedly impenetrable barriers and blithely materialize on the other side. And in the particle entanglement we explored in the previous chapter, *something*—some knowledge or influence or unknown entity—penetrates unlimited depths of space in zero time. All these suggest that space is a funny thing, with travel possibilities we are only beginning to understand.

We've come a long way since Galileo sent metal balls rolling down ramps. We've explored the speeds and vagaries of nearly every kind of object in all areas of nature. As for these superluminal galaxies exceeding the limits of our understanding, well, such off-the-scale speeds numb the mind today, but—count on it—they will mean something else to our grandchildren.

A sudden breeze blew the curtain on my office window inward, knocking over a vase overfilled with dried flowers. I swallowed an oath before it even moved past my lips, my eyes drawn to the blowing branches outside the window. Was that you, Torricelli, conjuring some sort of closing statement?

Silly thoughts. I shook them off.

After all, it's looking more and more like Aristotle and Alhazen were right: motion never began.

There can be no final curtain.

Acknowledgments

My thanks to Jane Weinberg for her invaluable help. And to my editors, John Parsley and Barbara Clark, who made everything better.

Table of Selected Natural Speeds

VERY SLOW (NOT VISUALLY DISCERNIBLE)

Stalactites	1 inch / 500 years
Tectonic plates	1–4 inches / year
Mountains	1/7 inch–2.4 inches / year
Sea level (twenty-first century)	2 inches / decade
Toenails	1/2 inch / year
Fingernails	1/8 inch / month
Hair	1/2 inch / month
Trees	1–2 inches / month
Fastest-growing plant (bamboo)	1 inch / hour
Bacteria (typical)	6 inches / hour
Germs (fastest)	1 foot / hour
Undisturbed airborne dust	1 inch / hour
Sperm	1 inch / 4 minutes
Snails (typical)	1 inch / 50 seconds

SLOW BUT VISIBLE

Snails (fastest)	40 feet / hour
Sloths	2 inches–1.7 feet / second
Ants	0.20 miles / hour
Giant tortoises	0.23 miles / hour

VISIBLE

Rivers	3 miles / hour
Human swimmer (fastest)	4–5 miles / hour
Drizzle (salt-grain-size rain)	4–5 miles / hour
Large raindrops (house-fly-size)	22 miles / hour
Cumulus clouds (typical)	20–30 miles / hour
Sharks	30 miles / hour
Greyhounds	45 miles / hour
Ocean waves	45 miles / hour
Fastest land animal (cheetah)	60–70 miles / hour
Large hailstones	105 miles / hour
Meteorite striking rooftop	250 miles / hour
Tsunami at sea	500 miles / hour

SUPERSONIC

Sound through air (thunder)	1/5 mile / second
Earthquake waves	5 miles / second
Earth around sun	18.5 miles / second
Meteoroid entering Earth's atmosphere	5–40 miles / second
Sun and Earth around galaxy center	140 miles /second
Solar wind particles	300 miles / second
Light through glass	139,600 miles / second
Light through space	186,282.4 miles / second

All are consensus or average values.

A Note on Accuracy and Choice of Units

In a number of cases, authoritative sources cite conflicting information. How fast does Mount Everest rise annually? Some say 3.9 inches, others say 0.15 inch. I have contacted three university geology professors and received conflicting information even from them! What's the top speed of a sloth? Seemingly reputable sources cite figures that range from one hundred feet a minute to five feet a minute. In such cases of wild disagreement I have included the range of accepted data. In others, where the disagreement was smaller, I have simply listed the average.

While nearly all the world, including the entire science community, exclusively employs metric units, this book mostly expresses itself in US or Imperial units. The choice was deliberate and the reason simple: for the vast majority of Americans as well as many in Great Britain, the content will be more meaningful if expressed in familiar terms. For example, when we reveal the speed of falling rain, few would find "9.8 meters per second" as clear and meaningful as "twenty-two miles per hour."

Notes

Chapter 1

The Growth of Nothingness

1. Edwin Hubble is credited with discovering the expansion of the universe in 1929, but there's never a mention of the woman who lurked behind the scenes. Henrietta Leavitt was a brilliant astronomer in the early 1900s, at a time when women were so discriminated against that the best she could do was make thirty cents an hour performing menial "computing" at the Harvard College Observatory. Nonetheless, she single-handedly discovered star families and found she could determine a star's absolute brightness from the colors it emitted. Hubble used her data and her method to calculate galaxy distances, which let him make the astonishing announcement that the universe is inflating like a balloon. Her credit? Nearly zero.

2. What might it mean if everything in the cosmos got larger simultaneously? Could you possibly perceive any change if your eyes, your body, the wavelengths of light, the room, the planet, and the whole universe suddenly tripled in size during the next few seconds? The answer is no. Such an alteration would be undetectable. Indeed, perhaps that is already continually happening. Maybe the cosmos keeps blowing up and shrinking down, so that a minute ago it was all the size of a single atom. The point is, "the universe is expanding" is meaningless unless it's merely some parts of the universe that grow larger while others stay the same. That would be the only sort of occurrence that could be detectable or even have logical meaning. And, indeed, that is exactly what is happening. Galaxies and their contents stay the same size, more or less, as do clusters of galaxies. Only the gap *between* galaxy clusters is growing.

3. Despite the degradation in conditions due to light pollution, the observatory on Mount Wilson remains in operation today.

4. Plans are well along for an astounding superinstrument, the twenty-four-meter (thousand–inch!) Giant Magellan Telescope, which would far

surpass everything else on the planet, although the European Space Agency is now planning a slightly larger telescope to stay one step ahead in bragging rights. The GMT's first mirror segment is already completed, as are the on-site ground-clearing operations. The construction of the whole telescope is slated to be finished in 2020. It is located on Cerro las Campanas (Campanas Hill), a mountain in Chile.

5. You'd think the farthest visible galaxies would look the smallest, as they float way off in the yawning distance. But when their light started on its journey to us some thirteen billion years ago, the universe was much tinier, and these galaxies were actually not greatly separated from us at that time. They were relatively nearby and hence looked larger. And even though eons have passed since their images began to travel our way, through space that continually stretches to make the trip surrealistically lengthier, those galaxies' apparent size remains the same. Bottom line: as their light arrives here, these galaxies look much larger than objects at their present great distance would logically suggest.

6. To the limit of what we can observe, the universe continually gets ten trillion cubic light-years larger each second.

To grasp the concept of the cosmos growing ten trillion cubic light-years larger each second necessitates appreciating what a "trillion" and a "cubic light-year" are. A trillion is a million millions. Despite the fact that we encounter this number from time to time (the US national debt was $14 trillion in 2012), it is staggeringly large. To merely count to a trillion, at the rate of rattling off five numbers a second, would require the time that has elapsed since the Pyramids were built. The cubic light-year is an equally mind-numbing concept as a measurement of volume. You'd have to picture a cube in which each dimension is one light-year. Each side would be as long as six million suns in a row, but keep in mind that the sun is itself one hundred times the width of the earth. Indeed, if one dropped Earths into a cubic light-year at the rate of a thousand a second, and began at the moment of the big bang, this colossal cube would not remotely be filled even today.

7. Actually, the nature and origin of consciousness—perception—is probably an even greater mystery than the source of the big bang or the makeup of dark energy. How *awareness* can possibly arise from chemical compounds or from colliding atoms is utterly baffling and defies even the most desperate guesses. But dark energy is right up there with the all-time greatest conundrums.

8. Though Greek and Roman scholars periodically cited Aristarchus during the next few centuries, virtually nothing is known about the very first person who said that our world moves. Nearly two millennia before Copernicus, Aristarchus of Samos alone declared that it makes more sense for Earth to orbit around the sun—the larger body—while spinning like a top than for absolutely everything in the cosmos to be whirling around us, even if either reality would produce the same observed visual effect: celestial objects crossing our sky. Sadly, his wisdom arrived at a bad time. Aristotle's Earth-centered doctrine was already spreading far and wide. In a flash of determination and rash budget busting, I decided to go to Samos to unearth facts about Aristarchus that I couldn't find in any domestic sourcebook or online. Thus in July of 2012 I traveled to that large island in the Aegean Sea. I hired an interpreter, interviewed the curator at the Samos archeological museum as well as dozens of other people, and tried to learn something new. I figured it would make a cool chapter. I was wrong. Although the odyssey started out hopefully enough—the Samos airport is called the Aristarchus airport—it turned out that nothing about his life, at least after his youth, is known even in Samos. It doesn't help that all but one of his books have failed to survive. A record heat spell, where it hit 105 degrees each day, was the only reward for my efforts. Aristarchus, who first revealed that our world spins while zooming through space, remains a cipher. And your author, who'd expected to write a chapter crammed with revelations, ended up with no more than this footnote.

Chapter 2
Slow as Molasses

1. An old joke involves two campers surprised by a bear. One starts running away at full speed, the other follows. When they meet up, panting, the second guy says, "Why did you start running? Did you really think you could outrun a bear?" To which the first camper replies with a shrug, "I didn't need to run faster than the bear. Just faster than you."

2. A frustrating experience for any researcher, editor, or author is finding definitive data about something that ought to be well established. Some authorities say the Himalayas are rising by as much as 2.4 inches a year. Others say one-seventh of an inch annually. You'd think this would have been settled by now, in our modern era of GPS. Suffice it to say that over the course of each of our lives, Everest grows either a foot taller or sixteen feet taller.

3. A research competition has existed for the past eighteen years between scientists at NIST, the National Institute of Science and Technology, and

Wolfgang Ketterle's lab on Massachusetts Avenue in Boston, at MIT. Each has successively outdone the other in periodically creating the coldest temperature ever.

4. For readers who'd like a more complete sense of being there during that famous cataclysm, here is a much fuller firsthand account from Pliny. This specific description became so widely read that in modern times volcanologists officially categorize all explosive volcanic eruptions as "Plinian events." Pliny wrote:

> Your request that I would send you an account of my uncle's death, in order to transmit a more exact relation of it to posterity, deserves my acknowledgments...And notwithstanding he perished by a misfortune, which, as it involved at the same time a most beautiful country in ruins, and destroyed so many populous cities, seems to promise him an everlasting remembrance...
>
> My uncle...was at that time with the fleet under his command at Misenum. On the 24th of August, about one in the afternoon, my mother desired him to observe a cloud which appeared of a very unusual size and shape. He had just taken a turn in the sun, and, after bathing himself in cold water, and making a light luncheon, gone back to his books: he immediately arose and went out upon a rising ground whence he might get a better sight of this very uncommon appearance. A cloud, from which mountain was uncertain, at this distance (but it was found afterwards to come from Mount Vesuvius), was ascending, the appearance of which I cannot give you a more exact description of than by likening it to that of a pine-tree, for it shot up to a great height in the form of a very tall trunk, which spread itself out at the top into a sort of branches; occasioned, I imagine, either by a sudden gust of air that impelled it, the force of which decreased as it advanced upwards, or the cloud itself, being pressed back again by its own weight, expanded in the manner I have mentioned; it appeared sometimes bright and sometimes dark and spotted, according as it was either more or less impregnated with earth and cinders. This phenomenon seemed to a man of such learning and research as my uncle extraordinary and worth further looking into...
> Meanwhile broad flames shone out in several places from Mount Vesuvius, which the darkness of the night contributed to render still brighter and clearer. But my uncle, in order to soothe the apprehensions of his friend, assured him it was only the burning of the villages,

which the country people had abandoned to the flames: after this he retired to rest, and it is most certain he was so little disquieted as to fall into a sound sleep: for his breathing, which, on account of his corpulence, was rather heavy and sonorous, was heard by the attendants outside. The court which led to his apartment being now almost filled with stones and ashes, if he had continued there any time longer, it would have been impossible for him to have made his way out... It was now day everywhere else, but there a deeper darkness prevailed than in the thickest night; which, however, was in some degree alleviated by torches and other lights of various kinds. They thought proper to go farther down upon the shore to see if they might safely put out to sea, but found the waves still running extremely high, and boisterous. There my uncle, laying himself down upon a sail-cloth, which was spread for him, called twice for some cold water, which he drank, when immediately the flames, preceded by a strong whiff of sulphur, dispersed the rest of the party, and obliged him to rise. He raised himself up with the assistance of two of his servants, and instantly fell down dead; suffocated, as I conjecture, by some gross and noxious vapour... [T]he third day after this melancholy accident, his body was found entire, and without any marks of violence upon it, in the dress in which he fell, and looking more like a man asleep than dead. During all this time my mother and I, who were at Misenum — but this has no connection with your history, and you did not desire any particulars besides those of my uncle's death; so I will end here, only adding that I have faithfully related to you what I was either an eye-witness of myself or received immediately after the accident happened, and before there was time to vary the truth...

Pliny also wrote, to Cornelius Tacitus:

There had been noticed for many days before a trembling of the earth, which did not alarm us much, as this is quite an ordinary occurrence in Campania; but it was so particularly violent that night that it not only shook but actually overturned, as it would seem, everything about us... Though it was now morning, the light was still exceedingly faint and doubtful; the buildings all around us tottered, and though we stood upon open ground, yet as the place was narrow and confined, there was no remaining without imminent danger: we therefore resolved to quit the town. A panic-stricken crowd followed us, and (as to a mind distracted with terror every suggestion seems more prudent than its own) pressed on us in dense array to drive us forward as we came out. Being at a convenient distance from the houses, we stood still, in the midst of a most dangerous and dreadful scene... The sea seemed to roll back upon itself, and to be driven from its banks by the convulsive motion of the earth; it is certain at least the shore was considerably enlarged, and several sea animals were left upon it. On the other side, a black and dreadful cloud, broken with rapid, zigzag flashes, revealed behind it variously shaped masses of flame: these last were like sheet-lightning, but much larger... Soon afterwards, the cloud began to descend, and cover the sea. It had already surrounded and concealed the island of Capreae and the promontory of Misenum... The ashes now began to fall upon us, though in no great quantity. I looked back; a dense dark mist seemed to be following us... It now grew rather lighter, which we imagined to be rather the forerunner of an approaching burst of flames (as in truth it was) than the return of day: however, the fire fell at a distance from us: then again we were immersed in thick darkness, and a heavy shower of ashes rained upon us, which we were obliged every now and then to stand up to shake off, otherwise we should have been crushed and buried in the heap... At last this dreadful darkness was dissipated by degrees, like a cloud or smoke; the real day returned, and even the sun shone out, though with a lurid light, like when an eclipse is coming on. Every object that presented itself to our eyes (which were extremely weakened) seemed changed, being covered deep with ashes as if with snow...

5. *Pyroclasts* are rock fragments fashioned in a volcanic explosion. The term *Peléan* comes from the famous Mount Pelée on Martinique, where the term *pyroclastic flow* was first scientifically defined after the great 1902 eruption.

Chapter 3
Runaway Poles

1. Here's a cool *Jeopardy!* "answer": "Istanbul is figuratively called the place where East meets West. But that is literally true only here." Question: "What are the geographic poles?" This is where east and west merge and simply vanish as separate entities.

2. The speed of sound has a caveat. It changes its velocity depending on temperature. Not pressure or altitude, just temperature. For example,

throughout this book the speed of sound is given as 768 miles per hour, but this is the case only at room temperature, or 20 degrees Celsius, or 68 degrees Fahrenheit. Sound moves significantly more slowly, at only 741 miles per hour, at the freezing point, 32 degrees Fahrenheit. It zooms 770 miles per hour at 72 degrees Fahrenheit, 773 miles per hour at 75 degrees Fahrenheit, and 776 miles per hour when it's 80 degrees Fahrenheit.

3. Although most animals are "blind" to Earth's magnetic field, behavioral studies have shown that some can indeed sense it. These include sea turtles, stingrays and manta rays, homing pigeons, migratory birds, honeybees, salmon, sharks, and tuna. Researchers have discovered that each of these creatures' nervous systems contains magnetite deposits. These small, naturally occurring magnetlike crystals align themselves with magnetic fields and act as microscopic compass needles. Such crystals are undoubtedly the key biologic component that allows some animals to sense Earth's magnetic field and navigate thereby.

4. Is it just me? Or is the term *Cretaceous Superchron* unbelievably cool-sounding? I've started to use it at every opportunity, even when it's not appropriate. People then ask me what it means, providing a reason to say it yet again.

Chapter 4
The Man Who Only Loved Sand

1. Dust devils form when hot air just above the surface quickly rises through a small region of cooler air just above it. In near-calm conditions, any horizontal motion begins the process of rotation. The fast-rising hot air pocket gets stretched vertically, shifting its mass closer to the axis of rotation, which intensifies the spin, following the law of conservation of angular momentum — just as a whirling ice-skater pulls in his arms to increase the spin. The rising hot air also creates a partial vacuum near the ground, pulling in other nearby hot air, which rapidly whooshes horizontally inward to the base of the whirlwind, adding to the spin. Thus the vortex is intensified and self-sustaining.

2. Later I learned that typical dust devil winds blow at forty-five miles per hour. Giant dust devils reach sixty miles per hour. The all-time record was seventy-five miles per hour. Looking back, I think that yes, I could have stepped into one without too much risk. However, I'm sure my attorney, if I had one, would insist that I unequivocally state that I am *not* suggesting you try it.

3. Well? How do the stars appear when viewed from an excellent, unpolluted earthly site as compared to the view from space itself or from the moon? I asked the man on earth who is arguably in the best position to know: Commander Andy Thomas, a longtime NASA astronaut who has logged months in space. He grew up in the Australian outback and knows what dark skies are, on earth and off it. He confirmed the science literature that says our atmosphere only dims stars by a barely noticeable one-third of a magnitude. In other words, there's a far bigger difference in star count when one moves from dark suburbs to even darker suburbs than there is when one blasts off into space and stargazes above our atmosphere. Air is very transparent to light's visual wavelengths.

4. Leap seconds are not fun for everyone. There is currently a raging debate over whether to do away with them once and for all and to change our clock-keeping system so that we no longer stay in sync with our planet's spin. Early in 2012, an international panel was so divided that they've shelved the issue until 2016, when it will be debated anew.

5. As for the other stages of twilight, *nautical twilight* persists until the sun is twelve degrees down. That's when the horizon vanishes; when a mariner cannot distinguish between sea and sky. *Astronomical twilight* continues still longer, until the sun has fallen eighteen degrees below the horizon, allowing the faintest stars to emerge. Its conclusion heralds the arrival of full darkness. The onset and duration of all three stages of twilight are not expressed in units of time but rather in terms of the Sun's distance below the horizon, because their lengths vary. Depending on the time of year and the latitude of the observer, twilight can expire in less than an hour or linger throughout the night. Twilight is always shortest in the tropics, where one hour of total twilight is all you get. From the latitude of New York, one and a half hours is about average, while from northern Europe, there simply is no night at all between May and August.

6. Singing sand is a real phenomenon, even if its cause remains a mystery. Obviously, motion is always necessary to create sound, but what exactly is the cause of singing sand, and what are the requisite preconditions? This latter question has been answered, because singing or booming only occurs when the sand grains are round, between 0.1 and 0.5 millimeters in diameter, at a specific humidity, and contain silicon dioxide (as sand usually does). The tone is commonly around the musical note of A, similar to a mosquito's drone, often with a deep undertone between 60 Hz and 105 Hz. It can be extremely

loud. And the phenomenon has been observed in dozens of deserts around the world.

Chapter 5
Down the Drain

1. Want to know how fast Earth whirls you in *your* hometown? It's easy with any scientific calculator. First punch in your home's latitude. (Don't know it? Just Google it. Brooklyn is 40° north; Seattle is 48° north.) Next, hit the COSINE key and you'll see a number between zero and one. In the case of Brooklyn, it's 0.766. Multiply this by 1,038 miles per hour, and you're done. Result: people in Coney Island go 795 miles per hour, just over the speed of sound, even when they're not riding the famous Cyclone roller coaster.

2. The green flash occurs when the last tiny spot of setting sun changes from orange to green, just for a second or two. You might see it once in about every sixteen or seventeen sunsets, if my experience is any guide to its frequency. I've seen it fifteen times, although I've looked for it about 250 times. It happens because the sun's image is actually composed of multiple colors that slightly overlap. When all the other "suns" have set, the topmost tip of the final one would be blue, except there isn't any blue light left, because it's been scattered away by the thick air at the horizon. So the actual topmost sun is green. But it's only seen when the air is very calm and homogeneous in temperature, as it sometimes is over the ocean.

3. If you approach a circular, counterclockwise-spinning storm from the outside, as incoming air does, the wind's deflection is toward the right. But if you're trapped inside the storm, as I was not too long ago, then the winds blow from right to left.

4. Going sixty-nine miles north from Quito produces a difference of a mere 0.2 miles per hour in Earth's rotation speed. But if you travel that same sixty-nine miles north from Point Barrow, Alaska, you come to a place where Earth spins a whopping seventeen miles per hour slower. Hence, ironically, despite all the tourist fuss paid to the Coriolis force at the equator, it is so negligible there that rotating storms like hurricanes can never form.

5. You'd think the rotating Earth would make a Foucault pendulum complete a 360-degree pivot after one period of Earth's rotation, which is twenty-three hours and fifty-six minutes. But, as it happens, this is only true for a pendulum at the North Pole or South Pole. Anywhere else the pivot

takes longer because the pendulum's direction precesses. This is a difficult math and physics conundrum ultimately attributable mostly to the Coriolis force, which initially baffled nineteenth-century physicists. Even Einstein found it puzzling enough to write about.

Chapter 6

Frozen

1. You don't always have to specify whether you're using Fahrenheit or Celsius during the Alaska winter. The two scales meet at negative forty degrees. During 2012 in Fairbanks the temperature hovered between minus forty and minus fifty degrees the entire month of January. Perhaps surprisingly, one can very much feel the difference between the two. While minus forty is bitter to the point of being surreal, it's actually dangerous to inhale air at minus fifty degrees because it can freeze lung tissue.

2. Snowflakes have not been extensively analyzed for the presence of germs. The issue was studied in France in 2008, when researchers found that 85 percent of flakes had formed around a living bacterium. Presumably this is true everywhere, but no one can say for sure whether any one country's snowfalls are more sanitary than another's.

3. In some coastal Alaskan towns, the melt problem is not underground but rather on the surface. Because of vanishing sea ice during the summer, ocean waves now crash into buildings instead of what used to be permanent buffers of ice. Their replacement in the summer by open water is creating community nightmares. The village of Kivalina, in northwestern Alaska, is one example. Its population of three hundred people faces relocation, with a price tag estimated at $54 million.

Chapter 7

April's Hidden Mysteries

1. This of course is why we put food into refrigerators. By merely dropping the temperature to thirty-nine degrees Fahrenheit, we put an enormous brake on the speed of myriad biological processes, including those needed by reproducing bacteria. Pull the plug and let the temperature jump a paltry twenty degrees, and the milk quickly sours, allowing the biology pageant to unfold anew.

2. Writers nowadays are cautioned against using the word *awesome* because of its clichéd omnipresence since the 1990s. At a grocery recently, the

clerk asked if I had the exact change; when I fished it out, he said, "Awesome."

"No," I replied. "The Grand Canyon is awesome. Exact change is not awesome." But the unfolding of spring? Absolutely: resurrect that adjective.

3. The hundredth monkey was a popular concept in the 1970s. The story is that a researcher observing monkeys on a tropical island saw one washing its food before eating it, to remove the sand. He noted that very soon after, other monkeys were doing the same — which was behavior never before seen in this type of simian. A sort of evolutionary action was apparently occurring.

Now comes the eerie part. Within a single year, researchers suddenly started seeing the same behavior among this species of monkey in other parts of the world. The conclusion was astounding: when some critically large number of animals begin thinking or acting in a particular way, the phenomenon reaches a sort of tipping point, and the idea pops into the minds of *all* those creatures simultaneously, no matter where in the world they may be.

It's the old ESP thing, never laid to rest by science. And while appropriately New Age in flavor for the time, it didn't seem totally outrageous. Flocks of birds and schools of fish seemingly make simultaneous turns, as if mentally connected.

Ken Keyes adopted the idea in his popular book, *The Hundredth Monkey.* Keyes suggested that if enough people participate in the peace and environmental movements, they will suddenly "take off" and become the established behavior of the entire human race.

A wonderfully utopian notion. But meanwhile what came to light was that the original story was fictitious. Turned out there never was a researcher who noticed increasing numbers of monkeys washing fruit. Zoologists noted that monkeys have always periodically washed fruit.

4. Actually, a dial tone consists of two notes. One is indeed a musical A, at 440 cycles per second. The other, a quieter undertone, hums at 350 cycles per second, which is an F. If you place a vibration-sensing guitar tuner against a phone, it will alternate between announcing it's detected an A and an F.

5. If you're a stickler for precision, the wavelength of aurora light is usually 557.7 nanometers while that of a firefly glow is between 561 and 570 nm. The green-yellow of both appear virtually identical, but the firefly is very slightly yellower.

Chapter 8

The Gang That Deciphered the Wind

1. In some sourcebooks, Australia's Barrow Island is given the prize for having the strongest-ever wind gust—253 miles per hour. It was recorded during tropical cyclone Olivia on April 10, 1996. It surpassed the longtime previous record of 231 miles per hour, set on Mount Washington on April 12, 1934. However, the Mount Washington record was for a normal day, not a cyclone occurrence, and in any case the New Hampshire mountain has much higher ongoing mean wind speeds. Thus it probably deserves to retain its "world's windiest" title.

2. This was a horrible crash. The experienced pilot was trying to give his passengers a scenic view of Japan's sacred mountain, and he had no warning that the winds were so terribly turbulent that day. The aircraft disintegrated and crashed, killing all 113 passengers and eleven crew members, including seventy-five Americans who worked for or were family members of people who worked for the Thermo King corporation of Minneapolis, Minnesota. The accident left sixty-three children orphaned. This—and some frightening experiences I went through during my two thousand hours as a pilot—makes me leery of mountain flying whenever the wind is higher than about thirty miles per hour. Over and around Mount Washington it's often three times that.

3. Here is heliocentrism again, centuries before Copernicus and Galileo.

4. There really is no magic number that pinpoints where our atmosphere ends, because at no point does the air terminate abruptly. Higher than fifty-two miles above the earth, however, so few atoms exist that sunlight is no longer measurably refracted. We see no detectable glow above that point, although some curious atmospheric phenomena still occur, such as the burning up of meteors (at between sixty and eighty miles) and the glow of the aurora (at between sixty and 120 miles). Even above twenty-three miles, the air is too thin to support any kind of specialized aircraft wings. Another thing to consider is that the sky at forty-nine miles is no longer dark cobalt blue but black.

5. The lowest sea-level barometric pressure ever recorded was 24.69 inches, or 870 millibars, in the eye of a typhoon, and the highest was 32.01 inches (1,084 millibars) in Siberia on an exceptionally cold day. More than enough to make your ears pop. Cold air is denser than warm air, and dry air

is denser than humid air. Thus cold, dry air has its molecules packed closest together. Altogether, this represents an impressive maximum sea-level variation in pressure of 20 percent. Experiencing that much of pressure change would normally require ascending or descending a vertical mile, as you would if you travel from New Orleans to Denver.

Chapter 9
Blown Away

1. Here is a summary of what you experience with each Saffir-Simpson category of hurricane. These descriptions are excerpted from those created by the the Saffir-Simpson team (Timothy Schott, Chris Landsea, Gene Hafele, Jeffrey Lorens, Arthur Taylor, Harvey Thurm, Bill Ward, Mark Willis, and Walt Zaleski) and are used with the kind permission of the National Oceanic and Atmospheric Administration.

Category 1 Hurricane (sustained winds 74–95 mph, or 119–153 km / hr)
Very dangerous winds will produce some damage. This means that older mobile homes could be destroyed and some poorly constructed frame homes can experience major damage, such as the loss of the roof covering and awnings. Masonry chimneys can be toppled. Even the best-made frame homes could have damage to roof shingles, vinyl siding, and gutters. Windows in high-rise buildings can be broken by flying debris. There will be occasional damage to commercial signage, fences, and canopies. Large tree branches will snap, and shallow rooted trees can be toppled. Extensive damage to power lines and poles will likely result in power outages that could last between a few and several days. Hurricane Dolly (2008) is an example of a hurricane that brought category 1 winds and damage to South Padre Island, Texas.

Category 2 Hurricane (sustained winds 96–110 mph, or 154–177 km / hr)
Extremely dangerous winds will cause extensive damage. There is a substantial risk of injury or death to people, livestock, and pets from flying and falling debris. Older mobile homes have a very high chance of being destroyed, and the flying debris generated can shred nearby mobile homes. Newer mobile homes can also be destroyed. Poorly constructed frame homes have a high chance of having their roof structures removed. Unprotected windows will have a high probability

of being broken by flying debris. Well-constructed frame homes could sustain major roof and siding damage. Failure of aluminum screened-in swimming pool enclosures will be common. There will be a substantial percentage of roof and siding damage to apartment buildings and industrial buildings. Unreinforced masonry walls can collapse. Windows in high-rise buildings can be broken by flying debris. Commercial signage, fences, and canopies will be damaged and often destroyed. Many shallowly rooted trees will be snapped or uprooted and block numerous roads. Near-total power loss is expected, with outages that could last from several days to weeks. Hurricane Frances (2004) is an example of a hurricane that brought category 2 winds and damage to coastal portions of Port Saint Lucie, Florida.

Category 3 Hurricane (sustained winds 111–129 mph, or 178–208 km / hr)

Devastating damage will occur. There is a high risk of injury or death to people, livestock, and pets from flying and falling debris. Nearly all pre-1994 mobile homes will be destroyed. Most newer mobile homes will sustain severe damage, with potential for complete roof failure and wall collapse. Poorly constructed frame homes can be destroyed by the removal of the roof and exterior walls. Unprotected windows will be broken by flying debris. Well-built frame homes can experience major damage involving the removal of roof decking and gable ends. There will be a high percentage of roof-covering and siding damage to apartment buildings and industrial buildings. Isolated structural damage to wood or steel framing can occur. Complete failure of older metal buildings is possible, and older unreinforced masonry buildings can collapse. Numerous windows will be blown out of high-rise buildings, resulting in falling glass, which will pose a threat for days to weeks after the storm. Most commercial signage, fences, and canopies will be destroyed. Many trees will be snapped or uprooted, blocking numerous roads. Electricity and water will be unavailable for between several days and a few weeks after the storm passes. Hurricane Ivan (2004) is an example of a hurricane that brought category 3 winds and damage to coastal portions of Gulf Shores, Alabama.

Category 4 Hurricane (sustained winds 130–156 mph, or 209–251 km / hr)

Catastrophic damage will occur. There is a very high risk of injury or death to people, livestock, and pets from flying and falling debris.

Nearly all pre-1994 mobile homes will be destroyed. A high percentage of newer mobile homes also will be destroyed. Poorly constructed homes can sustain complete collapse of all walls as well as the loss of the roof structure. Well-built homes also can sustain severe damage, with loss of most of the roof structure and/or some exterior walls. Extensive damage to roof coverings, windows, and doors will occur. Large amounts of wind-borne debris will be lofted into the air. Wind-borne debris will break most unprotected windows and penetrate some protected windows. There will be a high percentage of structural damage to the top floors of apartment buildings. Steel frames in older industrial buildings can collapse. There will be a high percentage of collapse to older unreinforced masonry buildings. Most windows will be blown out of high-rise buildings, resulting in falling glass. Nearly all commercial signage, fences, and canopies will be destroyed. Most trees will be snapped or uprooted and power poles downed. Fallen trees and power poles will isolate residential areas. Power outages will last for weeks to possibly months. Long-term water shortages will increase human suffering. Most of the area will be uninhabitable for weeks or months. Hurricane Charley (2004) is an example of a hurricane that brought category 4 winds and damage to coastal portions of Punta Gorda, Florida, with category 3 conditions experienced elsewhere in the city.

Category 5 Hurricane (sustained winds greater than 157 mph, or 252 km / hr)

Catastrophic damage will occur. People, livestock, and pets are at very high risk of injury or death from flying or falling debris, even if they're indoors in mobile homes or frame homes. Almost complete destruction of all mobile homes will occur, regardless of age or construction. A high percentage of frame homes will be destroyed, with total roof failure and wall collapse. Extensive damage to roof covers, windows, and doors will occur. Large amounts of wind-borne debris will be lofted into the air. Wind-borne debris damage will occur to nearly all unprotected windows and many protected windows. Significant damage to wood-roof commercial buildings will occur from the loss of roof sheathing. Complete collapse of many older metal buildings can occur. Most unreinforced masonry walls will fail, which can lead to the collapse of the buildings. A high percentage of industrial buildings and low-rise apartment buildings will be destroyed. Nearly all windows will be blown out of high-rise buildings, resulting in falling glass, which will pose a threat for days to weeks after the storm. Nearly all commercial signage, fences, and canopies will be destroyed. Nearly all trees will be snapped or uprooted and power poles downed. Fallen trees and power poles will isolate residential areas. Power outages will last for weeks to possibly months. Long-term water shortages will increase human suffering. Most of the area will be uninhabitable for weeks or months. Hurricane Andrew (1992) is an example of a hurricane that brought category 5 winds and damage to coastal portions of Cutler Ridge, Florida, with category 4 conditions experienced elsewhere in southern Miami-Dade County.

2. The critical fact of cloud birth is that cold air cannot hold as much moisture as warm air. The difference is dramatic. At one hundred degrees Fahrenheit, air can hold *ten times* more water than it can at thirty-two degrees Fahrenheit. So when warm air rises and cools, there comes a height where it's cooled to its water-holding limit. At that moment of saturation the invisible vapor turns into untold billions of tiny liquid droplets: a cloud. This is why clouds usually have flat bottoms. That's the altitude and temperature at which that day's air reaches its dewpoint. Drier air must rise farther in order to cool enough to be saturated, which explains why clouds are much higher on crisp days than on humid days.

3. An open secret in the forecasting business is that meteorologists *love* violent weather. This is when all their book training about low pressure and close-together isobars comes alive. A hint of this secret reached public awareness with Sebastian Junger's 1997 bestseller, *The Perfect Storm:* people realized that *perfect* had one meaning for meteorologists and the opposite meaning for everyone else.

Chapter 10
Falling

1. These speeds assume no air resistance, which adds a bit of imprecision to falling speeds because it varies according to how spread-out you are—e.g., whether you're plummeting in a dive or with limbs extended, as is taught in skydiving classes. With splayed arms and legs, a falling person travels at forty-two (rather than forty-four) miles per hour after two seconds and sixty (rather than sixty-six) miles per hour after three seconds.

2. The place in the sky around which all the constellations and stars pivot—similar to the stationary leg of the drafting compass we used at school

to draw circles — is called the North Celestial Pole. Polaris happens to sit less than one degree from that spot. But thanks to our planet's 25,780-year axis wobble, this stationary celestial point slowly shifts its location over the centuries and rarely happens to lie within one degree from any naked-eye star. At the time of the ancient Greeks, the star that most nearly didn't move had just changed from Thuban, in Draco, where the main passage in the Great Pyramid at Giza roughly points, to Kochab, in the Little Dipper. The current polestar, Polaris, is, by chance, the brightest star closest to the North Celestial Pole in the entire twenty-six-millennium precession cycle. Polaris doesn't seem to budge as the night wears on.

3. A veterinary study of cats that had fallen from high-rise buildings showed that 90 percent of them survived and that 30 percent of those that did had no injury. Mice and squirrels also have nonlethal terminal velocities; the fastest speed of a falling mouse would be just 1 percent of that of a falling elephant, according to physics (not actual experience). In fact, no fatal altitude exists for most small rodents: their terminal velocities are low enough to prevent acceleration to a lethal speed no matter what height they fall from. However, especially in cats, injury avoidance is aided by the ground often being a bit soft. It's not rocket science to conclude that it's better to land on a lawn than on a sidewalk.

4. The Greeks disbelieved in nothingness because they were such scrupulous logicians. What do we experience after death? To those who'd say, "We are nothing," they'd counter that the verb *to be* contradicts nothingness. To combine "is" or "are" with "nothing" is nonsensical. You can't "be nothing" any more than you can "walk not walk." Nothingness is a contradictory, meaningless concept — words without substance. You seem to be saying something, but you're not. By their reasoning, a vacuum cannot exist. Today we get their logic, it remains flawless, and yet they were wrong anyway. That's because the real world is not obligated to live by the rules of human language, which relies on symbolism. Actual water is not the word *water,* and the word *it* corresponds to nothing at all in the phrase "it is raining."

5. It remains little known and rarely discussed in Western classrooms today, but there is convincing evidence that ancient Indian astronomers beat out all the Renaissance scientists when it came to discovering gravity's existence. A full millennium before Newton, in the seventh century, Brahmagupta, living in Rajasthan, said, "Bodies fall towards the Earth as it is in the nature of the Earth to attract bodies, just as it is in the nature of water to

flow." Nor had he merely stumbled on such profundities by guesswork. He was a brilliant mathematician, the person who invented (or perhaps we should say *discovered*) the number zero.

Yet even he might not have been first. A century earlier, another Indian, named Varāhamihira, whom we discussed in chapter 8, wrote of a force that might be keeping everything stuck to the earth. This even went beyond the concept of local falling objects; Varāhamihira, critically, recognized that this force applies to the sun pulling on the planets. The very word for gravity in Sanskrit—coined centuries before Newton—is *gurutvakarshan,* which means "to be attracted."

6. Here's why an apple falling off a branch displays the same behavior as the moon. The moon is sixty times farther from Earth's center than the apple is, and thus it should experience 60 × 60, or 3,600, times less gravity than the apple. So instead of falling at the apple's rate of twenty-two miles per hour faster each second, it should fall 3,600 times less fast, or just 0.006 miles per hour—about six inches a minute. That's the speed of dust settling after you've shaken out a rug. Thus the moon barely falls. And, while it does, the moon also travels horizontally forward at the rate of 2,200 miles per hour. The two combined motions result in a curved path. The moon goes forward at just the correct speed so that our planet's curvature drops the ground from directly beneath it at the same rate, and thus it never gets close enough to us to experience a stronger gravitational pull; that's why it never gains speed. It travels ahead and also falls downward, maintaining the same distance, and therefore it orbits around us forever.

7. By assuming a streamlined diving position, a skydiver can attain a speed of two hundred miles per hour.

8. Galileo disproved the widespread belief that heavy objects fall faster than light ones. But when you trip and fall, *shouldn't* you be pulled downward more quickly than a lighter object? The surprising answer is: you are, even though it doesn't make you fall any faster. Heavy objects are indeed yanked more forcefully than light ones. Say you're the late world chess champion Aron Nimzowitsch, who actually once leaped on a chess table and shouted, "Why must I lose to this idiot?" If when he jumped off he simultaneously knocked a chess piece to the floor, they both hit the ground at the same time. Gravity pulled on his body with greater power than it tugged on the pawn. However, since the chess champion weighed so much, his mass took longer to speed up, just as a truck accelerates more sluggishly than a sports

car. The result is a wash. His body's yanked with more force, but it speeds up more reluctantly, and both objects fall at the same rate.

9. In particular, the direction in which a planet's lopsided elliptical orbit angles away from the sun is not fixed. The orbit itself twirls around like a squashed Hula-hoop, changing its orientation in space. Even the moon's oval orbit keeps changing the direction in which its longest dimension, which performs a complete rotation around Earth every 8.86 years, is aimed. So it's not just the moon that circles us; its elliptical orbit whirls around us as well, at a rate 118 times slower. The planet Mercury's squashed orbit does the same, but twice as quickly as can be explained by Newtonian physics.

10. Einstein even messed up his own calculations. He originally came up with a very wrong figure for the amount of spacetime distortion at the surface of the sun. That would have been disastrous for him, because the best test of his theory was to measure a star's position at the solar edge. Distant starlight skims right over the sun's limb, or edge, en route to our eyes, traversing the place of maximum spacetime bending. It should, according to Einstein, make the light take a longer path and cause the star to appear in an unexpected position—a deflection that, he said, should be readily measurable.

When can we see and measure a background star adjacent to the blinding solar edge? During a total eclipse! Thanks to World War I, a 1915 eclipse that could have tested the relativity theory was not suitable for viewing—an expedition to see it would have been unsafe. But by the time a total eclipse approached in May of 1919, an event in which the darkened sun would be fortuitously positioned amid the many stars of the Hyades cluster in Taurus, Einstein had corrected his math and offered a new figure for a star's expected deflection from its catalog position. Famed British astronomer Arthur Eddington, an Einstein booster, led an expedition that did indeed measure exactly the predicted results, which made Einstein a household name overnight. But skeptics howled. Eddington had used a tiny telescope with a four-inch mirror. The observations were performed in turbulent daytime air; the star images were blurry and dancing. The required accuracy was an arc second—the apparent size of a twenty-five-cent coin seen at a distance of three miles. Had Eddington really confirmed Einstein, or, rather, did he merely see what he wanted to see?

The famous 1919 results remain controversial to this day. No matter; later observations confirmed relativity. Spacetime was real. The motion of celestial objects was attributed to their journey across curved space.

11. Unlike the three other fundamental forces that describe relations between physical systems, gravity remains remains mysterious. The other three—electromagnetism (which manifests itself as magnetism and electric fields and such), weak nuclear force, and strong nuclear force, which operate only within the tiny regions in atoms—have even been theoretically tied together. But gravity eludes all attempts to weave it into any larger picture, to connect it with the others.

Chapter 11
Rush Hour for Every Body

1. Actually we must distinguish animal limbs from a weight or bob at the end of a wire, a true pendulum, to which the wire doesn't contribute very much to the total mass of the device. If instead a heavy, rigid rod is used in a pendulum—or, in this case, the rigid massive bone of the femur—then the oscillation rate matches that of a true pendulum (one on which the bob is nearly the entire mass and what holds it is a negligible mass, like a wire) that's two-thirds the length. So on a human, whose foot is just a smallish mass compared with the entire weight of the leg, that two-thirds business comes closest to providing us with the observed period. To use real numbers, a weight on a wire that's thirty-nine inches long will have a round-trip period of two seconds. But a human foot acting as a bob, at the end of a thirty-nine-inch bone, will swing as if it's on a pendulum just twenty inches long and complete the back-and-forth in about 1.5 seconds.

2. The first national highway using the Scotsman John Loudon McAdam's method was an eighty-foot-wide triumph that headed west from Cumberland, Maryland, which eventually became part of US Route 40. But McAdam's real contribution was in creating more cost-effective ways of building these roads—and of popularizing them. The three-layers-of-stones method, with the finest compacted at the top, had been designed decades earlier by the Frenchman Pierre-Marie-Jérôme Trésaguet. Perhaps it was simply easier on the tongue to call them *macadam* roads.

3. Want to know how fast you're likely to travel on your next trip? Here are more airliner speeds. The Boeing 777 goes 639 miles per hour, the 767 does 609 miles per hour, and a slew of commercial jets—including the Airbus A320, A310, and the omnipresent Boeing 737-800—fly at 594 miles per hour. If you're nostalgic for the old stagecoach experience, then travel on the older but still commonplace Boeing 737-300/400/500 models. They lope along at just 563 miles per hour.

4. What's the fastest any of our body parts can ever go naturally? It's a close call. There are only two contenders. The best pitchers can hurl a fastball at 102 miles per hour, which means that a pitcher's own fingertips are traveling through the air that quickly at the moment of release. This would match the fastest-ever sneeze on record, creating a tie for the "swiftest speed a person can achieve" award. (The absolute fastest pitch on record was thrown in 2010, when the twenty-two-year-old Cincinnati Reds left-handed reliever Aroldis Chapman made history by hurling the fastest pitch ever measured by radar in a major-league game, at 105 miles per hour.) Maybe there ought to be separate trophies for voluntary and reflexive categories.

Chapter 12

Brooks and Breakers

1. The aquatic ape hypothesis (AAH) states that many otherwise puzzling features of Homo sapiens can be explained *aquatically*. The hypothesis was first proposed in 1942 by German pathologist Max Westenhöfer and, independently, in 1960 by British marine biologist Alister Hardy. It was tirelessly publicized by the Welsh author Elaine Morgan in books such as *The Aquatic Ape* and *The Descent of the Child*.

We didn't start out on the savanna as just a smarter variety of ape, goes this line of reasoning. Instead, our ancestors were stranded (probably in East Africa) during a period of rising sea levels. Or, alternatively, we found ourselves with too much competition on land and took to finding our fortunes in lakes and inland waterways. Perhaps one large colony of our ancestors became marooned on an island at a time of rising sea levels and had to learn to live on the beach and make its fortune from the ocean.

Our ancestors started using tools because they needed to pry open clams and such. We spent more time in the sea and soon lost our furry coats, because hairlessness is an advantage in water. Perhaps the hair on our heads remained so that our young could have something to hold on to when we swam. Our noses grew far longer than that of chimps so we could breathe more easily when trying to keep our heads up. Our fat became attached below the skin, just like that of dolphins and whales, rather than forming as a separate layer, like that of all the other apes and land mammals.

When surprised or terrified, we gasp. Why? Apes never gasp. It only makes sense if it's the vestigial legacy of taking a sudden breath in order to dive.

The AAH also explains why we are so obsessed with water. Other apes will cross water only when they have to, or if there's food on an opposite bank. They don't love it. We do: we vacation at lakesides and by the sea, and a newborn baby will instinctively act appropriately and not drown—at least not right away—if thrown into water. (Don't try this!) While the aquatic ape hypothesis is largely ignored or belittled by paleoanthropologists, I wouldn't be surprised if schoolchildren a century hence are taught that explanation of our origins.

2. It's no accident that liquid water exists within a 180-degree range, from thirty-two degrees to 212 degrees. When Daniel Fahrenheit created his scale, he wanted ice and steam, which he regarded as opposite states of matter, to be represented by numbers that were opposites of each other. In a circle or compass, the opposite direction, an "about-face," is 180 degrees from where you started. Moreover, the geographic poles lie 180 degrees of latitude apart. The same goes for longitude: places farthest east or west from the zero point, at Greenwich, England, have longitudes of 180 degrees. So Fahrenheit made his scale's one-degree gradations of the necessary size so that 180 of them would mark the progression from freezing to boiling. As for why he chose such an odd number to be water's freezing point, it was because his zero was the coldest liquid he could produce—a slurry of near-frozen salt water. Starting there, he found that plain water freezes thirty-two degrees higher up.

3. Here is another top *Jeopardy!*-level fact: the moon is a place where water could freeze and boil *simultaneously*.

4. Accidents no longer rank among the top three causes of death, except among young people, but there is a major gender gap in accidental death rates. Only 3.5 percent of women die from an unintentional injury, but for men the rate is 6.5 percent. I doubt this will surprise anyone.

5. I just said "three-foot" tidal bulge located roughly beneath the moon, but coastal areas get an average five-foot tidal range, thanks to the amplification effects of the shallower seabed there. But out in the open sea it's three feet.

6. Where I live, one hundred miles up the Hudson River from New York City, the ocean tides manage to march all the way at full force. Once high tide hits Manhattan it progresses upriver at seventeen miles per hour and takes six hours to reach us, which means when it is high tide up here it is exactly the next low tide down in the big city. While the swell of the tide moves at seventeen miles per hour, the water itself does not. Anyone watching floating debris on Hudson River tides, as in that tidal bore, will see it progress

only slowly northward with the incoming tide and will then later observe it flowing south. It may repeat the zigzag process several times before it finally clears one's location for keeps. This is why it would take someone in an inner tube 126 days to float from my region to lower Manhattan—about four months to go that hundred miles. Few commuters choose this inexpensive travel option.

7. Compared to the historically grievous Alexandria tsunami, perhaps ten times more people died on the day after Christmas in 2004 as a result of the intense Indonesian tsunami, which was caused by a quake that released the power of 23,000 Hiroshima-type atomic bombs. But how long will it be memorialized? Do we annually think about those 228,000 people even now, a few years later, after they succumbed to the sixth-worst natural disaster in human history? This surely speaks of the big-heartedness of the Alexandrians, who kept alive the memory of the victims of the 365 CE cataclysm for more than two hundred years.

Chapter 13
Invisible Companions

1. Researchers finally realized that most of the radiation from radium took the form of alpha particles. These are heavy clumps of two protons and two neutrons, slow-moving enough that they cannot even penetrate skin. The real problem arises if radium is inhaled or swallowed. Then it is absorbed by bones, where its alpha emissions steadily bombard and destroy the marrow.

2. In 1938, after winning the Nobel Prize, Hess relocated to the United States with his Jewish wife to avoid Nazi persecution. He immediately became a Fordham University physics professor and lived in an apartment in Mount Vernon, New York, until his death in 1964.

3. Physicists call ordinary light *electromagnetic radiation,* yet it's not the least bit harmful; we're not zapped with danger whenever we flip a wall switch. Long-wave light, like the kind we see, and the infrared rays we can feel on our skin can't damage atoms. Visible light, radio waves, Wi-Fi, even microwaves cannot abuse genes and cause cancer. Living next to a cell phone tower means that invisible microwaves are flying through you at light speed and can make entire atoms in your body jiggle. This can slightly heat up tissue, but it can't break those atoms or make tumors form. (It still might not be good for you!) By contrast, gamma rays' and X-rays' short waves do break apart atoms. This is "ionizing radiation," the bad kind that can sabotage

301

genes. Heavy, fast-moving subatomic particles, such as neutrons and protons, can also destroy atoms, so they're often called radiation even though they're particles and not energy waves. The distinction is blurry anyway, since all matter has a wavelike aspect.

4. One trustworthy reference source says that a one-light-year-thick lead wall is excessive. That you could stop the average neutrino with a mere body of water that stretched from the sun to Saturn. That's 6,400 times narrower than a light-year and would be a lot cheaper to obtain, even if you couldn't order it on Amazon. Either way, catching neutrinos isn't something you do casually.

5. Life's general rule is that moving entities are often endearing as long as they can't attack us and as long as there's not a lot of them. Two birds chirping outside the window is a joy. A flock of two thousand birds would be something out of a Hitchcock flick. One ladybug on the windowsill is pretty. The presence of a thousand requires remedial action. Squirrels, chipmunks, ants, even intestinal flora: we want our in-motion companions to present themselves in limited quantities. Yet neutrinos and probably dark matter constantly pervade us in unimaginable numbers, but they don't have "bad PR" because they're not just invisible: *they also do not meddle.*

Chapter 14
The Stop-Action Murderer

1. The hummingbird's dual nature—now you see it, now you don't—combined with its exquisite colors made it an obsession for some early civilizations. The Aztec god Huitzilopochtli is usually depicted as a hummingbird. And the outline of one appears in the famous Nazca lines of Peru.

2. By playing their mythology card, the Greeks created imaginary ways creatures could move. Having the head of a human and body of a horse, for example, yielded a composite that somehow offered only the advantages of each. My favorite such hybrid creature was envisioned by Woody Allen. It had the head of a lion and the body of a lion—a *different* lion.

3. The former chairman of Columbia University's astrophysics department, David Helfand, told me that he has used the waving fingers technique to "freeze" and observe the famous pulsar at the heart of the Crab nebula. It blinks on and off thirty times a second, far beyond a person's ability to perceive it as anything but steady light. Yet Helfand made it blink by rapidly

oscillating his hand—making a motion like that of fan blades—in front of the telescope's eyepiece. Low tech, but it worked.

4. Modern movies display three of the same frame in a row, with moments of darkness between, then three images of the next frame, and so on, for a total of seventy-two images along with seventy-two momentary periods of darkness per second—even if there are only twenty-four *different* images presented each second. When this method is used, no one complains of seeing any flickering.

5. Iron rusting is a leisurely process because it requires fast-moving colliding atoms. In ordinary real-life situations, the average speed of iron atoms is too leisurely for reactions with oxygen to proceed. But at any given moment, a few atoms move faster than the group average, and it is these that continually create the oxidizing.

6. If increasingly heated metal glows red, then orange, yellow, and white, and if blue would be the next color if the substance didn't first boil into gas, what happened to green? All the rainbow colors are represented except for green. Why? The answer, which also explains why there are no green stars, is that when green light is being maximally emitted the human eye sees it as white. That's because, at that point, there are still copious red and blue emissions in the mix, and our retinas perceive white whenever those three primary hues hit us simultaneously. White *is* our green, in these cases.

Chapter 15
Barriers of Light and Sound

1. One's home is not always a safe house when it comes to lightning. I once carefully assembled firsthand accounts for a safety article published in 1984. The first was from a friend whose extended family gathered for Thanksgiving in Catskill, New York. Through the bay window, several of them saw lightning strike a large tree at the far end of the lawn. Immediately, a "ball" of lightning appeared at the base of the tree. It started to "roll" along the grass toward the window, coming right toward them. It briefly vanished from sight below the glass, but then, to their horror, all the seams in the drywall started to glow. Suddenly the dazzling ball was inside the wall and resumed its "rolling" across the living room. My friend said that his elderly aunt who hadn't walked for years leaped out of its way as it went directly into the TV set, which exploded in a shower of sparks. For many long seconds, all was silent. Then his characteristically laconic father finally spoke. "I guess that takes care of *that* TV," he said.

My second story involves a woman in the village of Saugerties, New York. This was a well-known incident in 1983. She said it happened on a day with a clear blue sky, before any storm could be seen. She was in her house when lightning explosively struck the roof, flinging a rain of asphalt shingles down the block. She was struck in the head in her living room, and the electricity exited her big toe, leaving a black burn. Though many of her teeth were shattered and she required months of rehabilitative care, she attributed her survival to the fact that she'd been wearing rubber flip-flops. I asked her if she was now afraid of lightning. "No; of course not!" she assured me. "It was a one-in-a-million event. I merely take the same precautions everyone does. I make sure I'm wearing flip-flops all the time, no matter what."

2. None of the other sense-propagation velocities mattered. Or even earned attention. Few wondered how fast *smells* travel. (Actually, we have—in chapter 7. Along with the speed of neural impulses conveying the senses of touch and of pain, in chapter 11.)

3. When Galileo telescopically observed Saturn between 1610 and the 1630s, he described that planet—in words and pictures—as having handles on either side, like those of a sugar bowl. It took until Christiaan Huygens's later observations, a full half century after Galileo's, for the rings' true nature to come to light. Why? Probably because here on earth there is not a single example of a ball surrounded by unattached rings. It lay outside human experience. Observers had a hard time seeing something that was without precedent. The same impediment may have prevented anyone from regarding lightning as preceding thunder. In the prefirecracker era, no one knew of any lights that produced sounds. Lightning would have been the first to do so.

4. Through air, sound moves only one way—by compressing and then decompressing the gas. In effect, it pushes along a disturbance in the air, which dampens down in time, thus explaining why sounds get fainter and less distinct as the distance increases. Sound's so-called longitudinal waves, moving only in the direction of travel, are also present when sound goes through solids. However, in the latter case, a second wave exists, too. This is the up-and-down, or elastic, deformation of the material, usually called a shear wave or transverse wave, and it can actually travel at a different speed from the longitudinal wave—letting the listener receive two separate noises. The speed of shear waves in solids was calculated accurately by Isaac Newton in his all-purpose 1687 masterpiece, *Principia*. The speed is determined by the solid's density, stiffness, and susceptibility to compression.

5. The wave-versus-particle furor reminds me of the old joke about the agreeable judge who never wanted to make anyone feel bad. After one side argued its case in his courtroom, he said, "You're right." Then the opposing side made strong antithetical arguments, to which the judge said, "*You're* right!" Hearing this, the first plaintiff rose with exasperation and said, "But Your Honor, we've made opposing points. We can't both be right!" The judge just smiled and said, "You're right!" In the same way, the wave and particle evangelists are all correct.

6. Actually, for us to see white, the mix only has to include equal amounts of the primary colors—red, green, and blue. Unequal mixtures of any two or all three of these can create every other imaginable color. But the primary colors of paint and pigment are cyan, magenta, and yellow. Artists create other colors by mixing these, but unlike light, which requires only that more light of different wavelengths be added in order to change color, paint requires the *subtraction* of some of the light being reflected from the mixture in order to change color. A canvas doesn't glow on its own. Rather, a picture is viewed in white light, and each of its pigments absorbs one or more colors present in the room's light, so that what reflects to your eyes is the hue the artist wanted that spot to be. Thus adding further pigments subtracts more of the ambient light. In point of fact, each primary color of paint is composed of an equal mixture of two of light's primary colors. That is, red and green light combine to create yellow light, and yellow is a primary color of paint. Similarly, red and blue light make magenta. Blue and green make cyan.

7. Could we have any advance warning of something arriving at light speed? No. In *Star Wars*–type movies, the hero's spaceship skillfully dodges and weaves to avoid laser weapons and photon torpedoes. In reality there's no way to anticipate the arrival of a light-based weapon's pulses or rays, no way to "see them coming." However, we would be able to detect *reflections*. So say the sun suddenly went dark. Although we wouldn't see this happening ahead of time, we could see the various planets blink off one by one as light reflected from their surfaces no longer arrived. Mercury would vanish first, then Venus. Saturn would keep shining for more than an hour after Earth's sunlit hemisphere lost its light. Thus if the sun's demise occurred at night, we'd have advance notice of it without having to wait for the sunrise that never arrives.

8. If this doesn't yet seem bizarre, imagine if baseballs behaved as photons do. Imagine driving a pickup truck at ninety miles per hour directly toward a

batter while a pitcher standing in the flatbed hurled his best one-hundred-mile-per-hour fastball. The ball should logically reach the batter at an unhittable 190 miles per hour. But what if the ball arrived at the strike zone at the same speed regardless of the vehicle's motion, even if it were speeding *away* from the plate? Would that not be odd? Yet that is exactly how photons behave.

9. To ponder more logical behavior, consider sound waves. When we approach the source of sound, as when a wailing ambulance is racing toward us, its siren's waves hit us at a faster speed. This scrunches them up, and the pitch audibly rises. It's the famous Doppler shift. But when we approach a light source, its waves do scrunch up to change the observed color (since blue light waves are closer together than red ones), yet the speed of each photon never budges. This is bizarre and counterintuitive.

Chapter 16
Meteor in the Kitchen

1. The 1908 Tunguska intruder is usually characterized as a meteor. This is merely the generic term for any object that arrives on Earth from space. An asteroid made of metallic rock and a comet made largely of ices are both termed a meteor or meteorite when, respectively, they're zooming across the sky or hitting the ground. There is a small-minority view that the Tunguska event (and, for that matter, the "great dying," the Permian extinction event that took place 251 million years ago) was caused by trapped gas escaping from deep within the earth and then igniting high in the air. But the vast majority of scientists are confident that it was an air-bursting meteor that did not survive its passage through the atmosphere. This would also explain the absence of any crater or meteoric debris. Moreover, air-bursting meteors have been chronicled in the past, whereas an escaping glob of methane that did not explode until it rose miles into the atmosphere would be a unique event in world history.

2. NASA and the Russians also left spent landers on Venus, Mars, and the Saturnian moon Titan. A probe also parachuted into Jupiter, but it was swallowed up and crushed by the planet's thick gases, so we won't count that as an example of littering, because it could be argued that the probe is "out of sight, out of mind."

3. For millennia, one of the permanent frustrations for astronomers was the inability to observe the moon's hidden side. Yet everyone expected it to

look more or less like the face we do see. That's why the Russian *Luna 3* probe, which whizzed past the far side in October of 1959, its television cameras whirring, created such a shock. The hidden hemisphere was a different world! It had virtually none of the large, dark blotches—the so-called seas—that give the familiar side its characteristic, chauvinistically named man-in-the-moon appearance. Obviously that more distant portion escaped the period of volcanism the near side underwent. This is supported by the fact that the moon's center of mass is not in its geographical middle but rather a mile closer to Earth. Meanwhile, the Russians exercised their prerogative as discoverers and gave Russian names to every mountain and crater and almost every pebble, an embarrassment that has kept that lunar hemisphere out of many Western textbooks.

Chapter 17

Infinite Speed

1. Actually, FitzGerald couldn't believe that light is always a constant regardless of one's motion toward or away from the light source. He assumed that observers and their measuring tools had their length squashed in the direction of travel in such a way that light would merely *appear* constant. He thought that fast speed introduced an experimental distortion.

Chapter 18

Sleepy Village in an Exploding Universe

1. Because the speed of the expanding universe increases with distance, weird stuff happens to truly faraway objects. Consider a galaxy at the edge of the visible universe. We can say it's old because we see it as it was when its light started traveling to us thirteen billion years ago. Its image is ancient. We can also say it's young because we're viewing a picture of a newborn galaxy; after all, everything back then was newly hatched. But is it really thirteen billion light-years away, as news articles claim? Does it even make sense to compare where we are now with where that galaxy was situated thirteen billion years ago? When the image we're seeing left that galaxy, we were much closer together. It was then only 3.35 billion light-years from us. So it should logically display the size of a galaxy at that nearer distance—its location when its light left—rather than the size of a galaxy located as far away as it is now. A photograph's dimensions don't change just because it took a long time to get delivered.

Amazingly, that galaxy indeed looks much larger than we'd expect for something so far away. It's like a fun-house mirror. The galaxy *appears* much closer than it is!

In terms of its size, that is. But it's far dimmer than we'd expect an object at that distance to be. Space has been stretching while the image traveled, dramatically redshifting and weakening it. It now exhibits the ultrafaintness of a galaxy at the impossible distance of 263 billion light-years.

Let's put all this together. It's the oldest galaxy image we've ever seen, but it's of a newborn galaxy, so we can also say it's the youngest. It looks way too big for its distance, but also way too faint. Could things get any weirder? You bet. Science articles say it's thirteen billion light-years from here because distance is often expressed that way—as how many light-years of space the image had to traverse to arrive here. And thirteen billion years is also how long its light took to reach us. However, during all that time the galaxy has meanwhile been madly receding. This galaxy is now *actually* thirty billion light-years away. Its recession speed today is far faster than that of light.

2. Every astronomer in the early 1990s would have told you with absolute certainty that the universe's expansion is slowing down. There was even a name for this: *the deceleration parameter*. But just a few years later, the cosmos had flipped. Then it was obvious that the expansion is *speeding up*. Cosmologically we are clearly in our infancy. Despite TV specials carrying on about the overall universe being such and such, the scarcity of hard data makes virtually nothing we "know" immune from revision and reversal tomorrow. Those knowledgeable in astrophysics greet all the popular speculative models with a smile.

Bibliography

The data in this book come from hundreds of sources. For example, a single sentence about the respective growth rates of willow trees and maple trees comes from a poster for homeowners published by a Florida utility company, which obtained the information from the Arbor Day Foundation. For this bibliography, the twenty-one data resources listed below contain trustworthy, meaty content for follow-up explorations.

Books

Bagnold, R. A. *The Physics of Blown Sand and Desert Dunes.* Mineola, N.Y.: Dover Publications, 2005.

Bova, Ben. *The Story of Light.* Naperville, Ill.: Sourcebooks, 2001.

Considine, Glenn D., ed. *Van Nostrand's Scientific Encyclopedia.* 9th ed. 2 vols. Hoboken, N.J.: Wiley-Interscience, 2002.

Gosnell, Mariana. *Ice: The Nature, the History, and the Uses of an Astonishing Substance.* New York: Alfred A. Knopf, 2005.

Leonardo da Vinci. *The Notebooks of Leonardo da Vinci.* Edited by Edward MacCurdy. Old Saybrook, Conn.: Konecky & Konecky, 2003.

McLeish, Kenneth. *Aristotle.* New York: Routledge, 1999.

Meeus, Jean. *Astronomical Tables of the Sun, Moon, and Planets.* 2nd ed. Richmond, Va.: Willmann-Bell, 1995.

Pliny the Younger. *Letters.* Translated by William Melmoth. Revised by F. C. T. Bosanquet. Harvard Classics vol. 9, part 4. New York: P. F. Collier & Son, 1909–14.

Weisberg, Joseph S. *Meteorology: The Earth and Its Weather.* 2nd ed. Boston: Houghton Mifflin, 1981.

Websites

Casio Computer Co., Ltd. Keisan Online Calculator. http://keisan.casio.com/has10/Menu.cgi?path=06000000.Science&charset=utf-8.

Darling, David. The Encyclopedia of Science. http://www.daviddarling.info/encyclopedia/ETEmain.html.

Elert, Glenn, ed. The Physics Factbook: An Encyclopedia of Scientific Essays. http://hypertextbook.com/facts/.

Goklany, Indur M. "Death and Death Rates Due to Extreme Weather Events: Global and U.S. Trends, 1900–2004." Center for Science and Technology Policy Research, University of Colorado at Boulder. http://cstpr.colorado.edu/sparc/research/projects/extreme_events/munich_workshop/goklany.pdf.

Heidorn, Keith C. "The Weather Legacy of Admiral Sir Francis Beaufort." http://www.islandnet.com/~see/weather/history/beaufort.htm.

Heron, Melonie. "Deaths: Leading Causes for 2008." National Vital Statistics Reports 60, no. 6 (June 6, 2012). United States Department of Health and Human Services, Centers for Disease Control and Prevention. http://www.cdc.gov/nchs/data/nvsr/nvsr60/nvsr60_06.pdf.

Laird, W. R. "Renaissance Mechanics and the New Science of Motion." Canary Islands Ministry of Education, Universities, and Sustainability. http://www.gobiernodecanarias.org/educacion/3/usrn/fundoro/archivos%20adjuntos/publicaciones/largo_campo/cap_02_06_Laird.pdf.

Llinás, Rodolfo. "The Electric Brain." Interview with Rodolfo Llinás conducted by Lauren Aguirre for Nova online. http://www.pbs.org/wgbh/nova/body/electric-brain.html.

National Weather Service National Hurricane Center. Saffir-Simpson Hurricane Wind Scale. http://www.nhc.noaa.gov/aboutsshws.php.

Nave, C. R. HyperPhysics. http://hyperphysics.phy-astr.gsu.edu/hbase/hph.html.

Sachs, Joe. "Aristotle: Motion and Its Place in Nature." Internet Encyclopedia of Philosophy. http://www.iep.utm.edu/aris-mot/.

Sengpiel, Eberhard. "Calculation of the Speed of Sound in Air and the Effective Temperature." http://www.sengpielaudio.com/calculator-speedsound.htm.

Index

About the Author

BOB BERMAN, one of America's top astronomy writers, contributed the popular "Night Watchman" column to *Discover* for seventeen years. He is the author of *The Sun's Heartbeat* and is currently a columnist for *Astronomy,* a host on Northeast Public Radio, and the science editor of the *Old Farmer's Almanac.* He lives in Willow, New York.